DECISION MAKING FOR ENERGY FUTURES

DECISION MAKING FOR ENERGY FUTURES

A Case Study of the Windscale Inquiry

David Pearce, Lynne Edwards and
Geoff Beuret

First published 1979 by
THE MACMILLAN PRESS LTD
London and Basingstoke
Associated companies in Delhi
Dublin Hong Kong Johannesburg Lagos
Melbourne New York Singapore Tokyo

British Library Cataloguing in Publication Data

Pearce, David William
   Decision making for energy futures
   1. Atomic energy – Great Britain
   2. Energy policy – Great Britain
   3. Decision-making
   I. Title II. Edwards, Lynne
   III. Beuret, Geoff
   IV. Social Science Research Council *(Great Britain)*
   621.48′0941   TK9057

ISBN 978-1-349-04987-5   ISBN 978-1-349-04985-1 (eBook)
DOI 10.1007/978-1-349-04985-1

Essentially, the challenge facing policy makers is how to devise acceptable and practical methods of reaching collective decisions for a technology that involves risks that some people find unacceptable, dissenting minorities who oppose the siting of nuclear plants in their locality, poor understanding of the technical issues among the public and disagreement among technical experts. Clearly, there are no panaceas, for it is a uniquely difficult problem to adapt decision making procedures to accommodate a wide range of viewpoints on the technical, environmental and political aspects of nuclear power.

John Surrey and Charlotte Huggett

'Opposition to Nuclear Power: a Review of International Experience', *Energy Policy*, December 1976

Contents

Note

This report was commissioned by the Energy Panel of the SSRC from Professor D. W. Pearce with the aim of identifying what we could learn from the proceedings of the Windscale Inquiry that might help the conduct of future public discussion on energy matters. The report is not intended to reflect the views of the Energy Panel, but the Panel believes it is a most useful contribution to public debate.

The Energy Panel is an SSRC initiative whose aim is to promote and encourage social scientific and inter-disciplinary research into matters connected with energy. Its members (all of whom serve in a personal capacity, and not as representatives of their institutions) are:

Michael Posner (Chairman)
Social Science Research Council,
Fellow of Pembroke College, Cambridge

Professor P. M. Abell
Birmingham University

F. Colmer
Central Electricity Generating Board

Professor Ian Fells
Newcastle University

The Lord Flowers, F.R.S.
Imperial College of Science
& Technology

Professor David Henderson
Department of Political Economy
University College London

T. A. Kennedy
Department of Energy

J. V. Mitchell
British Petroleum Co. Ltd.

Professor P. Odell
Erasmus University

M. J. Parker
National Coal Board

Professor D. W. Pearce
Aberdeen University

I. Smart
Independent consultant

A. J. Surrey
Science Policy Research
Unit, Sussex University

Corresponding member
R. L. Nicholson
United Kingdom Atomic
Energy Authority

Dr. B. D. Jones
Science Research Coun-
cil Assessor

Acknowledgements

Countless individuals have helped us in our work and we wish to thank them all for giving freely of their time and for reading various papers and chapters in draft form. It should be made absolutely clear that none of them bears any responsibility for the views expressed in this report. If there are errors of fact or interpretation, they are our own. We do not, however, make any apologies for the views expressed here. They are our own and we stand by them.

We owe particular debts of gratitude to British Nuclear Fuels Limited at Risley, Cheshire, for allowing us to scrutinise their records of the Windscale Inquiry, for discussions, exchanges of written views, and technical advice. We are indebted to the Atomic Energy Authority in London and especially to Len Brookes for discussions about the likely future of inquiries into nuclear-related investments. We would also like to thank the Director of the Dounreay Reactor Establishment for a stimulating visit to the United Kingdom's only fast reactor and for discussions with himself and his staff. The Departments of Energy and Environment offered comment on the Interim Report produced in May 1978, for which we are grateful. We have received invaluable legal advice from Clive Brand of Aberdeen University Law Department; from Oliver Thorold and Geoffrey Searle of the Lawyers' Ecology Group, Dr J. N. Hawke of Leicester Polytechnic, and Jeremy Rowan-Robinson of the Department of Land Economy, University of Aberdeen. We have benefited from planning advice from Professor Alistair MacLeary of Aberdeen University and Peter McVean, now of PEIDA, Edinburgh, and from discussions on planning matters with David Hall and his staff at the Town and Country Planning Association. Valuable advice was also received from the Department of the Environment on planning issues. Local residents near Windscale expressed their views through personal interviews at a local meeting with one of the researchers and in written statements. Journalists Ian Breach and Anthony Tucker offered valuable help in some areas, while numerous discussions took place with the so-called 'cause' groups over the year. If we did not manage to consult all of them this was because of time constraints. For comments on various sections we would like to thank Czech Conroy and Tom Burke of Friends of the Earth; Brian Wynne of the School of Independent Studies at Lancaster University; Gerald Leach of the International Institute for Environment and Development; Sir Kelvin Spencer; Robin Grove-White of CPRE; Bill Badger, Chris Haworth and other members of West Cumbria FOE; Edward Acland and other members of NNC; Irene Coates of the Conservation Society; Edward Dawson of CoEnCo; Dave Elliott of SERA; Peter Taylor of PERG; Sheila Oakes of NIN and Peter Chapman of the Open University.

We greatly appreciate exchanges of information with Professor Edward Radford of Pittsburgh, Dr Ian Bowen of Woods Hole Oceanographic Institute, and Dr Ivan Tolstoy. The presence of USA experts at a UK public inquiry was unusual and the impressions of these 'visitors' to the UK system provided much food for thought. David Fishlock of the *Financial Times* put us right on some aspects of media coverage and Terence Price of the Uranium Institute offered a demanding forum for trying out some of our ideas at the Institute's annual conference in 1978. Professor David Henderson read the entire script and made detailed and invaluable comments for which we are truly grateful. If we have not always heeded his advice it has usually been because we feel some of the issues he raised require detailed assessment in work that we plan for the future. We are particularly conscious of the lack of national comparative analysis, a whole area of study in itself. It remains the case that we feel the manner in which a UK inquiry process was undertaken is an area of invaluable study in itself.

We owe an enormous debt of gratitude to the SSRC Energy Panel for funding the work on which this book is based. Particular debts are owed to Christina Hadjimatheou, Cathy Cunningham and the Chairman of the Energy Panel and of the SSRC itself, Michael Posner. They are of course exempted from all responsibility in the same way as everyone else.

Last, but very far from least, we owe great debts to wives and husbands who bore the brunt of the domestic upset that such apparently sensitive research seems to entail. Apologies are owed to British Airways for unsound travelling capacity on the part of one researcher, and to the National Health Service for excessive demands on their time by another. To Mrs Winnie Sinclair and Mrs Grace McGregor we can only express the sincerest of thanks for typing endless scripts, papers, letters, secret and not-so-secret notes, dealing with endless inquiries and for typing both the Interim Report and this Final Report. They achieved all this with typical Scots humour and amidst a haze of incinerated tobacco that must have placed them, and us, well above Mr Justice Parker's acceptable level of risk in a city already built of granite containing uranium.

D. W. P.
L. E.
G. B.

Glossary of Abbreviations and Technical Terms

ABBREVIATIONS

AEA	*See* UKAEA
AGR	Advanced Gas-cooled Reactor
AWRE	Atomic Weapons Research Establishment
BEIR	Biological Effects of Ionizing Radiation (independent US Committee on effects of radiation)
BNA	British Nuclear Associates
BNOC	British National Oil Corporation
BNFL	British Nuclear Fuels Ltd
BWR	Boiling Water Reactor
CANDU	Canadian Deuterium-moderated Natural Uranium-fuelled Reactor
CDFR	Commercial Demonstration Fast Reactor
CEE	Commission on Energy and the Environment
CEGB	Central Electricity Generating Board
CoEnCo	Committee for Environmental Conservation
CENTEC	Centrifuge Technology Co.
CFR	Commercial Fast Reactor
COGEMA	Compagnie Général des Matières Nucléaires
CPRE	Council for the Protection of Rural England
DAP	Decision Advisory Procedure
DMP	Decision-Making Procedure
DFR	Dounreay Fast Reactor
EC	Energy Commission
EEC	European Economic Community
EPA	Environmental Protection Agency (USA)
EPC	Energy Policy Commission
ERDA	Energy Research and Development Association
EURATOM	European Atomic Energy Community
FBR	Fast Breeder Reactor

FOE	Friends of the Earth
GDP	Gross Domestic Product
GNP	Gross National Product
GW	gigawatts (million kilowatts)
GW(e)	gigawatts electric
HALW	High-Active Liquid Waste
HARVEST	Highly Active Residues Vitrification Engineering Studies
HASW	High-Active Solid Waste
HAW	High-Active Waste
HWR	Heavy Water Reactor
IAEA	International Atomic Energy Agency
ICRP	International Commission on Radiological Protection
INFCE(P)	International Nuclear Fuel Cycle Evaluation (Programme)
KEWA	Kernbrennstoff – Wiederanfarbeitungs – Gessellschaft mbH
LEG	Lawyers' Ecology Group
LALW	Low-Active Liquid Waste
LASW	Low-Active Solid Waste
LPI	Local Public Inquiry
LNG	Liquid Natural Gas
LWR	Light Water Reactor
MALW	Medium-Active Liquid Waste
MASW	Medium-Active Solid Waste
Mtce	Million tons (tonnes) coal equivalent
MW	megawatts (thousand kilowatts)
MW(e)	megawatts electric
NCB	National Coal Board
NCRP	National Committee on Radiation Protection (USA)
NII	Nuclear Installations Inspectorate
NIN	Nuclear Information Network
NNC	National Nuclear Corporation
NNC	Network for Nuclear Concern
NPT	(Nuclear) Non-Proliferation Treaty
NPV	Net Present Value – the current value of a future stream of costs and benefits discounted at some rate of interest
NRC	Nuclear Regulatory Commission
NRDA	Natural Resource Defence Agency
NRPB	National Radiological Protection Board
PERG	Political Ecology Research Group

PIC	Planning Inquiry Commission
PWR	Pressurised Water Reactor
RCEP	Royal Commission on Environmental Pollution
SDO	Special Development Order
SERA	Socialist Environment and Resources Association
SGHWR	Steam-Generating Heavy Water Reactor
SoSE	Secretary of State for the Environment
SSEB	South of Scotland Electricity Board
TCPA	Town and Country Planning Association
THORP	Thermal Oxide Reprocessing Plant
TNPG	The Nuclear Power Group
TRC	The Radiochemical Centre Ltd
UKAEA	United Kingdom Atomic Energy Authority
URENCO	Uranium Enrichment Company
WA	Windscale Appeal
WARP	Windscale Assessment and Review Project
WPI	Windscale Public Inquiry

TECHNICAL TERMS

Actinides	Elements following actinium in the periodic table. They include thorium, protactinium, neptunium, plutonium, americium, curium, berkelium and californium. Many of them are long-lived α-emitters.
Breed	To form fissile nuclei, usually as a result of neutron capture, possibly followed by radioactive decay.
Breeder	A reactor which can produce more fissile nuclei than it consumes.
Burn-up	Irradiation of nuclear fuel by neutrons in a reactor. It is measured in units of megawatts – days (of heat) per tonne of uranium or plutonium.
C	Caesium
CO_2	Carbon dioxide
Caesium	Particularly Caesium – 137. A fission product and biologically hazardous β-emitter.
Cave	A working space for the manipulation of highly radioactive items; it is surrounded by great thicknesses of concrete or other shielding and has deep protective windows.
Cladding	Material used to cover nuclear fuel in order to protect it

	and to contain the fission products formed during irradiation.
Coolant	Liquid (water, molten metal) or gas (carbon dioxide, helium, air) pumped through reactor core to remove heat generated in the core.
Cooling Pond	A deep tank of water into which irradiated fuel is discharged upon removal from a reactor, there to remain until shipped for reprocessing.
Core	The central region of a reactor where the nuclear chain reaction takes place, and heat is thereby generated.
Critical	Of an assembly of nuclear materials, that it is just capable of supporting a nuclear chain-reaction.
Decanning	Removal of cladding from fuel.
Decay	Disintegration of a nucleus through the emission of radioactivity.
Decay heat	Heat generated by the radioactivity of the fission products, which continues even after the chain reaction in a reactor has been stopped.
Depleted	Of uranium whose Uranium 235 content is less than the 0.7 per cent that tends to occur naturally.
Deuterium	Hydrogen 2, heavy hydrogen; its nucleus consists of one proton plus one neutron, rather than the one proton only of ordinary hydrogen.
Dose	The amount of energy delivered to a unit mass of a material by radiation travelling through it.
Enrichment	The process of increasing the concentration of the Uranium 235 isotope in uranium beyond 0.7 per cent in order to make fuel made from it more suitable for use in a reactor.
Fast	Of neutrons, that they are travelling with a speed close to that at which they were ejected from the fissioning nucleus.
Fast reactor	A reactor in which there is no moderator and in which the nuclear chain is sustained by fast neutrons alone.
Fertile	Of material like Uranium 238 or Thorium 232, which can by neutron absorption be transformed into fissible material.
Fissile	Of a nucleus, that it will fission readily if it is struck by and captures a neutron.
Fission	The splitting of a heavy nucleus into two (or more) parts, usually accompanied by a release of energy.

Fission
product — A nucleus of intermediate size formed from the breakdown or fission of a heavy nucleus such as that of uranium. Such a nucleus will be highly radioactive and usually emits β particles.

Fuel — Material (such as natural or enriched uranium or uranium and/or plutonium dioxide) containing fissile nuclei, fabricated into a suitable form for use in a reactor.

Fuel assembly, fuel element — A single unit of fuel plus cladding which can be individually inserted into or removed from the reactor core.

Fuel pin — A single tube of cladding filled with pellets of fuel.

Graphite — A black compacted crystalline carbon, used as neutron moderator and reflector in reactor cores.

Half-life — The time in which the number of nuclei of a particular type is reduced by radioactive decay to one-half.

Heavy water — Water in which the hydrogen atoms all consist of deuterium, the heavy stable isotope which is present to the extent of 150 parts per million in ordinary hydrogen.

Helium — A light, chemically inert gas used as coolant in high temperature reactors.

Hex — Uranium hexafluoride, a corrosive gas (above 56°C).

Hot-cell — *See* Cave.

Hulls — Fuel elements or parts thereof from which the uranium has been dissolved out by acid, leaving only the cladding.

Irradiated — Of reactor fuel, having been involved in a chain reaction, and having thereby accumulated fission products; in any application, exposed to radiation.

Isotopes — Two nuclei of the same chemical element that differ only in their mass, e.g. Plutonium 239.

Krypton — A chemically inert gas; the isotope Krypton 85 is a fission product at present released to the atmosphere from reprocessing plants.

Light water — Ordinary water, used as moderator and coolant in light water reactors.

Magnox — A magnesium alloy used as fuel cladding in the first-generation British gas-cooled reactors, which are therefore called magnox reactors.

Mixed oxide — A mixture of plutonium and uranium dioxides, used as the fuel in fast reactors.

Millirem (mrem)	1/1000th of a rem.
Moderator	A substance used to slow down neutrons emitted during nuclear fission.
Neutron	An uncharged particle, constituent of nucleus – ejected at high energy during fission, capable of being absorbed in another nucleus and bringing about further fission or radioactive behaviour.
Nuclear fuel cycle	The sequence of operations in which uranium is mined, fabricated into fuel, irradiated in a reactor, and either processed or stored.
Nuclear park	A large nuclear site in which there would be several large reactors, together with their associated fuel fabrication and reprocessing plants.
Nuclide	Any particular type of nucleus, not necessarily radioactive.
Pu	Plutonium.
Plutonium	A heavy 'artificial' metal, made by neutron bombardment of uranium; fissile, highly reactive chemically, extremely toxic α emitter.
Rad	Radiation absorbed dose; one rad equals the amount of radiation that will cause one kilogramme of material to absorb 0.01 joules of energy.
Radiation, nuclear	Neutrons, α or β particles or γ rays which radiate out from radioactive substances.
Radioactivity	The behaviour of a substance in which nuclei are undergoing transformation and emitting radiation.
Radionucleides	Nuclei that are radioactive.
Reactivity	Of fuel, its ability to support a nuclear chain reaction; it is a function of the concentration of fissile atoms and inversely of the quantity of neutron-absorbing material present.
Rem	Roentgen equivalent man: unit of radiation exposure, compensated to allow for extra biological damage by alpha particles or fast neutrons.
Reprocessing	The chemical separation of irradiated nuclear fuel into uranium, plutonium, and radioactive waste (mainly fission products).
SO_2	Sulphur dioxide.
Shielding	Material interposed between a source of radioactivity and an operator in order to reduce his radiation dose.

Spent   A term applied to fuel when it has been irradiated in a reactor for up to three years and is reaching the stage of diminishing efficiency.

Spiking   The irradiation of fuel with the intent of making it difficult or impossible to divert from legitimate use.

Tailings   Crushed uranium ore from which the uranium has been extracted chemically.

Thermal   Of neutrons, that they are travelling with a speed comparable with that of gas molecules at ordinary temperatures.

Thorium   A fertile heavy metal.

Tritium   Hydrogen 3—nucleus contains one proton plus two neutrons; radioactive.

Turnkey   A term applied to a system of contracting, whereby a single contractor is responsible for the whole of the design and construction of an installation.

U   Uranium

Uranium   The heaviest natural element, dark grey metal; isotopes 233 and 235 are fissile, 238 fertile; α emitter.

Vitrification   The incorporation of high-level wastes (mainly the oxides of metals formed as fission products) into a glass, ceramic or rock-like solid.

Yellow-cake   A mixture of the two oxides of uranium; a yellow powder.

Zircaloy   An alloy of zirconium used as fuel cladding.

OTHER

α(alpha) particle   A heavy, positively-charged particle; the nucleus of a Helium 4 atom containing two protons and two neutrons.

β(beta) particle   An electron; a light, negatively-charged particle.

γ(gamma) radiation   Electro-magnetic radiation of very short wavelength.

α-emitter   A radioisotope emitting α-particles.

β-emitter   A radioisotope emitting β-particles.

γ-emitter   A radioisotope emitting γ-radiation.

1 Introduction

In October 1977 the Energy Panel of the Social Science Research Council requested Professor Pearce of the Department of Political Economy at the University of Aberdeen to undertake a study of the implications of the Windscale Inquiry for the future of advisory procedures in the UK in the context of energy planning. The Windscale Inquiry was a local public inquiry, held at Whitehaven in Cumbria, into the proposal by British Nuclear Fuels Limited (BNFL) to build a plant to reprocess thermal oxide fuel from the UK's 'second-generation' nuclear reactors. It was proposed that the plant, known as THORP (thermal oxide reprocessing plant), be built at Windscale, the site of an existing nuclear reactor and of a plant to reprocess MAGNOX fuels (from the UK's first generation of reactors). For reasons considered in this report, the planning application by BNFL to Cumbria County Council was 'called in' by the Secretary of State for the Environment, Peter Shore, thus forcing a public inquiry into the proposal. That inquiry lasted 100 days, opening on 14 June 1977 and closing on 4 November 1977. In the following January, the Inspector, the Hon. Mr Justice Parker, presented his Report to Mr Shore. In March 1978 the Report was published. It proved to be a thorough vindication of BNFL's case, but occasioned severely critical comment from those who had opposed the building of THORP. Much of that criticism was directed at the way in which evidence had been treated by the Inspector. It led to calls for some new form of inquiry procedure, although, surprisingly perhaps, very few people suggested removing such inquiries from the planning law context altogether. The Report was greeted enthusiastically by BNFL and by the nuclear industry in general, who felt that many myths had been dispersed and that the categorical nature of the Inspector's judgements was the kind of assessment needed in the emotive and highly charged atmosphere surrounding nuclear power issues.

The current study is to be seen against this background. First, we did not begin work until the Inquiry itself had ended, although one of the authors had been attending the Inquiry as part of his own research work. While this may occasion criticism in itself, it was largely a matter of timing – the idea for the study did not emerge until the Inquiry was nearly ended. It was, however, in our view also a happy chance. While the Inspector could not have doubted that he was 'under study' from various quarters during the Inquiry, most of those studies have come from parties committed to particular views which were expressed at the Inquiry. One exception is Ian Breach's book *Windscale Fallout* (Penguin, 1978) which offers an informed journalist's view of the proceedings. To have announced a study by independent academics as the Inquiry was in process would not, in our view, have aided our research.

Second, events have moved rapidly since the Inquiry ended. The Inspector's Report occasioned more comment than the Inquiry itself. It was debated in Parliament in March of 1978 and again, briefly, in May of 1978. The recommendation to proceed with THORP was the subject of a rally in Trafalgar Square and at the time of writing is still being hotly debated. There is therefore some difficulty in gaining perspective on a sequence of events which has not ended, not least because of a promise by the United Kingdom Government, given in 1977, that the proposal to build the UK's first commercial-scale fast reactor would itself be the subject of some 'special' advisory procedure, to take place in the early 1980s.

Third, we have found that many issues relating to the way in which planning inquiries 'should' proceed are themselves debated. Any one inquiry may proceed according to laid-down rules of guidance to inspectors, and the inspector's own assessment of precedent and rules of procedure. No one inquiry therefore need proceed exactly like any other. This needs to be borne in mind when asking the question what the Windscale Public Inquiry (WPI) tells us about the general form that other inquiries should take.

Fourth, our concern throughout is with the 'efficiency' of inquiry procedures. To this end, we are not directly concerned with the actual arguments presented at Windscale. Clearly, if there was something in the format of the inquiry or in the nature of the inquiry processes themselves that led to evidence being misconstrued or relevant evidence being omitted, then the dividing line between the efficiency of *procedure* and the correctness of the final *advice* given by the Inspector will be a thin one. If, on the other hand, certain minimum standards are observed, it would be our view that sound advice may none the less emerge from an inefficient process although this in no way condones such processes. We would stress this argument since it is more than evident that both 'sides' in the nuclear debate will read into an assessment of the kind we present what they want out of it. This was all too painfully obvious to us after publication of our Interim Report. In suggesting that the WPI did not meet conditions that we would regard as satisfactory for full or overall efficiency, we are not suggesting that the Inspector proffered the wrong advice. Neither are we suggesting that he reached the 'right' decision. We divorce our terms of reference from the issue of whether or not spent nuclear fuels should be reprocessed in the UK. We fully understand the temptation to draw a logical inference from the efficiency of procedures to the 'correctness' of outcome. We suggest that this temptation should be resisted *unless* it is judged that the information presented at the inquiry was inadequate for sound advice to be given, or unless it can be demonstrated that the Inspector and his assessors had preconceived views of the proposal which should have been removed by the evidence presented.

The suggestions in this report are for improvements to the decision advisory procedures embodied at the moment in the form of public inquiries, Royal Commissions, Parliamentary Select Committees and certain standing commissions such as the Energy Commission and the Commission on Energy and the Environment. These are *advisory* institutions: they cannot themselves make decisions. Decisions lie with a Minister of State or with Parliament. Many other investments, the majority, do not even reach the stage of any inquiry process other than internal debate within a planning committee and a local council meeting. Quite why some issues need to be taken out of this local context is explained in the text. Suffice it to say that, while there is no hard and fast dividing line, there are national aspects to many investments and they require wider debate and discussion than is afforded by a local council meeting. For these 'national' issues we propose an expanded framework for assessment. In doing so, we are very conscious that, for many, we have suggested procedures which will appear to some people to 'play up' an issue and exaggerate its importance.

For others, our proposed procedures will not give *enough* attention to public debate and widespread consultation. We explain in the text why we have taken the views we have. We have not the slightest doubt that they will elicit some legitimate expressions of disagreement, and others which we see as reflecting the personal interests of various parties to the energy debate rather than the public interest as a whole. It is necessary to say this because it would seem that any work on issues related to nuclear power brings out reactions on the part of some which must be regarded as irrational. We have not been too surprised to learn that researchers elsewhere have experienced similar reactions, in some cases leading to their abandonment of the study of nuclear issues. Nor do we confine this judgement to either side of the debate. There are eminently sane persons in the anti-nuclear movement who genuinely believe there is a strong case against the development of nuclear power, a case based on fears about radiation hazards, nuclear weapons proliferation and civil liberties. Whether the balance of argument is in their favour or not is not our concern here. It is our concern that we see these objections as being *legitimate*, as being worthy of serious consideration, debate and investigation. That there are others in the anti-nuclear movement who have other motives for participation we have no doubt either. Equally, we have met and benefited greatly from those on the pro-nuclear side who genuinely believe in their industry and what they regard as its essential role in the future of an 'acceptable' society. They are prepared to admit fallibility and error, unproven technologies, and the fact that many arguments rest on a balance of pros and cons rather than a 100 per cent 'correctness' of view. We have also listened to pro-nuclear views that deny credibility in that they do not even admit of the possibility of error, and, while no doubt genuinely believed, are positively dangerous in ascribing a homogeneous political cause – the downfall of capitalistic societies – to the anti-nuclear movement.

There are therefore legitimate arguments on both sides, and, sadly, far too many views that do not withstand serious scrutiny. It is partly for this reason that we have suggested wider debate and informational procedures than some would wish to countenance. This also needs to be borne in mind when reading the document. We have spoken throughout of 'pro' and 'anti' views and we recognise a school of thought that regrets this continued polarisation. Rightly or wrongly, however, that is how the debate has developed in the UK and we are not of the view that it will change. This leaves one unfortunate aspect of accepting the existence of polarisation. This is that attempts to argue that things are not 'quite right' or 'could be better' are themselves seen as confirming one or other side in its views. In our own case, we have witnessed attempts by self-professed pro-nuclear persons to suggest that, because we advocate what we see as improvement of the decision advisory process, we are 'therefore' against nuclear power. Such incapacity for pursuing logical argument would not even deserve mention if we had not come across it so often. It is all too reminiscent of 'those who are not for us are against us'. Again, it is partly experiences such as this, again shared by other researchers, that lead us to urge a level of debate which deals with such fundamental issues as values and even the public image of the two sides of the debate.

As the text will make clear, it is not sensible to excise one part of a decision-making procedure, an advisory part constituted by a public inquiry, and look at it in isolation. Our criteria for 'efficient' decision procedures cover an *entire procedure*, from an application to build some installation through to the final decision. We have done our best therefore to isolate those suggestions which relate to the inquiry procedure alone, but much of what we have to say covers the entire process. This again needs to be borne in mind when reading the report. Finally, we have tended to concentrate on nuclear power. We have done this because it is difficult to envisage government policy in the UK doing anything other than espousing nuclear power as a means of sustaining a growth society. That is, nuclear power is fundamental to energy policy as the present or any likely British Government sees it. Whether this should be so or not is technically outside our terms of reference. None the less we must offer some discussion of the issue since we argue that the 'values' underlying the debate about whether nuclear power is needed or not are, in considerable part, what constitutes the debate. Failure to accommodate them in some way is a failure to proceed in an efficient procedural manner. This much said, the reader should not have too much difficulty in translating what is said of nuclear power to, say, policies for coal extraction. We give reasons for having 'picked on' nuclear power. Many of those reasons are, however, equally applicable to some other investments. The other reason for looking at nuclear power is self-evident. We were asked to look at the Windscale Inquiry, which was itself concerned with thermal oxide fuel reprocessing. What one does with spent fuel is itself fundamental to the choice of nuclear fuel cycle. We also give reasons in the text for suggesting that THORP is an integral part of *future* energy policy. Indeed, we accept that a nuclear future is not probable without THORP or what we would guess to be further reprocessing plant constructed towards the beginning of the next century. This does *not* mean we accept THORP – the judgement is a reflection of what we think is likely and not an assessment of the continuing debate on reprocessing versus storage. In looking at the Windscale Inquiry, then, we argue that we are considering decision-making for an energy future.

In considering the entire decision-making process we are of course aware that this includes the role of Parliament. It seems to us that Parliament has in the past been at some disadvantage when facing decisions on technological, especially perhaps 'high'-technology, issues because of the relatively few scientists and technologists of high calibre who have been attracted into government or the civil service. This makes the quality of the advice available to Parliament on such matters all the more important. If our proposals imply modification of existing Parliamentary procedures this might be thought an acceptable price to pay for better advice.

2 UK Energy Policy

INTRODUCTION

In February 1978 the Secretary of State for Energy presented his document *Energy Policy* (Cmnd 7101, 1978) to Parliament. Some of its contents had been released earlier under the aegis of the Energy Commission, a body composed in the main of the fuel industries' leaders and trades unionists (see Appendix 5). The Commission had been established in 1977 and met for the first time at the end of November 1977. As a 'consultative' document, *Energy Policy* is not supposed to indicate a commitment to any specific combination of energy sources or any firm faith in a set of energy forecasts. Indeed, policy is subject to a programme of fairly continuous review from within the Department of Energy, the constituent corporate bodies in the fuel industry (British Gas Corporation, the various Electricity Boards, the National Coal Board, the Atomic Energy Authority and the British National Oil Corporation), and the Energy Commission itself. Moreover, *Energy Policy* stresses the need for a 'flexible strategy', one which can be changed to suit changed circumstances.

All this said, no energy policy is capable of rapid reversal *in substance*, since the lead-times involved in the investment decisions are such that, once made, decisions commit the supply system some ten years or so ahead. There is, of course, room for flexibility within the context of the major programme decisions. Longer-term decisions are of course capable of adjustment and the Green Paper itself does not venture beyond the year 2000. The Energy Commission paper that preceded it (Energy Commission, 1977) did go beyond 2000 to 2025. In terms of *policy* the cut-off point at 2000 would seem eminently reasonable, not least because thereafter margins of error in demand and supply forecasts would probably swamp assessments of extra demand or supply. In terms of *research and development*, however, forecasting cannot stop at 2000 since research now is required to develop the energy sources of the twenty-first century. It is not possible, therefore, to avoid taking *some* view of prospective demand and supply after 2000.

7

THE OBJECTIVES OF ENERGY POLICY

As set out in *Energy Policy*, the objectives of energy planning are

(i) that energy supplies should not be a constraint on some predetermined economic growth objective;
(ii) that energy supplies to achieve this objective should be introduced so as to minimise the resource cost of so doing;
(iii) that some balance, expressible through the rate of discount, should be struck between the 'needs' of future and current generations;
(iv) that the minimum resource-cost objective should be extended to include the *use* of energy;
(v) that, to some extent, policy should aim to reduce import dependence.

As expressed, these objectives are themselves fairly 'elastic' in the sense that they do not always imply specific rules of magnitudes. Thus, the target rate of economic growth is whatever is thought to be desirable in the future. Energy policy must therefore be capable of responding to growth rates that may vary. To all intents and purposes, however, the assumed rate of growth of GDP has been taken to be a trend rate of $2\frac{3}{4}$ per cent to the year 2000. In so far as consideration is made for the years after 2000 the rate drops fairly significantly (see Department of Energy, 1978).

The minimum-cost approach is also ambiguous in that what is generally being referred to is resource costs. While the Green Paper discusses environmental costs and other external effects, it is not clear how far these costs are to be reflected in the minimum-cost formula. Clearly, whatever environmental standards are imposed on the various energy sources, either from within the UK or from without (e.g. the EEC), will impose costs on the industries and these costs will effectively become 'internalised'. Whatever environmental policy is pursued will therefore have its costs reflected in the minimum-cost approach. But clearly no view has been taken of the actual magnitudes of these costs, nor, given the state of applied economic science, does it seem likely that any such assessment could be made. That is, money values for environmental pollution simply cannot be calculated with anything other than spurious accuracy (Pearce, 1978).

None the less, it is precisely the approach to external costs that defines the differences between advocates of any given energy policy and the opposition. In terms of the current study, it is these differences of view that require debate, a debate which is promised fairly frequently in the Green Paper itself (or a debate which it is assumed is actually taking place). To take a simple example, if one believed that a fast breeder reactor had associated risks of terrorist or state proliferation of nuclear weapons using civil fuels, then there is some risk of these agents using those weapons if they secured them. The cost of their use could be regarded as infinite (what exactly is the cost of a world war?) by many. If so, multiplying an infinite cost by a probability of proliferation and a further probability of use of such weapons leaves us with an infinite external cost to be debited to fast reactors. Ergo, on this view, they would never be introduced. Others may think a world war or a limited war has a finite cost, so that a finite addition would need to be made to secure the 'right' relative cost structure between fast reactors and, say, thermal reactors, or coal, or whatever. Similar arguments could be advanced for CO_2 accumulation in the atmosphere, and so on. Given that no accurate money measures will exist for external costs, it follows that it is a matter of *judgement* as to what those external costs are. The impact of differing judgements would be to secure possibly very different energy policies. Indeed, exactly this explains why, say, the policy as outlined in the Green Paper looks so different from the policies that might be outlined by conservationist, environmentalist or 'cause' groups.

This issue is complicated by the possibility that any given energy policy may not itself be capable of sustaining a given rate of economic growth. If, say, policy A could secure a growth rate of 3 per cent but with an environmental cost of X (which would have to be an implied figure) and policy B could secure only 1 per cent with an environmental cost of Y where Y is less than X, then what is being 'traded' is 2 per cent economic growth for the magnitude $X - Y$. The 2 per cent economic growth can be translated into absolute money figures, bearing in mind of course all the reservations expressed about the use of GDP as a measure of the 'value' of goods supplied to society. In this way the absolute money figure can be compared to $X - Y$. If, as we argued above will be the case, all we have is some ordinal ranking of X and Y, then we at least have some approximation of the 'cost' (in terms of foregone GDP) of a given energy policy compared to another.

Of course, nothing is likely to be as simple as this. The issue will be complicated by uncertainty about energy supplies – a low-growth energy policy will have a premium attached to it in terms of the security of depletable fuels. But perhaps it will also be inconsistent with technological change, so that potential energy supplies are not developed, and so on. None the less, the basic comparison remains and, as a simple model, it is essential in evaluating energy policies. It is, of course, predicated on the idea that GDP and energy consumption are functionally related. There is no need to believe in any 'fixed' energy coefficient to justify this view. Indeed, the energy coefficient has changed substantially in the twentieth century in the UK in a downwards direction (a given percentage rate of growth of GDP is associated with a smaller and smaller percentage change in the rate of growth of energy consumption). The issue is whether it can change much further. Conservation measures are surely likely to secure once-for-all changes, and the use of renewable energy sources (e.g. solar power) can make some impact. Unless, however, one believes in the picture of a society in which all energy demands are met from renewable 'free' sources at no cost, then GDP and energy consumption must have some linkage. The degree of this linkage is debated. There are those who believe that little further progress can be made in 'decoupling' the two (Brookes, 1972) and those who believe that a total decoupling can (and should) occur (Lovins, 1977).

The main point being made here is that the acceptability of any energy policy will depend on (*a*) the empirical relationship between GDP and energy; (*b*) the extent to which economic growth is itself regarded as a socially desirable end and, if so, to what level; and (*c*) *judgemental* assessments of the external costs of different energy supplies. This is the context in which energy policy should be debated.

Finally, the Green Paper speaks of a 'balance' between current and future supplies, given the requisite demands of current and future generations. It argues that a high discount rate biases exploitation towards current generations and vice versa. The fact that the official government test discount rate of 10 per cent has been replaced by a wider and looser criterion of 5 per cent rate of return on overall investment is perhaps some slight indication of a change which, potentially, 'favours' future generations. Even then, a 7 per cent discount rate, say, implies that £1 in 50 years' time would actually only be recorded as three pence in present value terms. (At 10 per cent it would have been less than one penny.) But for any real recognition of the claims of future generations one would expect to see a depletion policy applied to the most immediately depletable fuels within the UK, oil and gas. It would take a stretch of the imagination to suggest that any such policy exists other than in respect of the most marginal impacts.

Indeed, one Department of Energy document states: ' . . . the import of oil is a powerful influence on economic policy through the balance of payments. The depletion of North Sea oil is therefore seen as a macroeconomic rather than an energy policy issue' (Department of Energy, 1978). Again, if there is to be an energy debate and if any sense is to be attached to the espoused concept of 'balance' it would seem right that depletion policies be far more openly countenanced and discussed.

THE FORECASTS

Table 2.1 summarises the forecasts as they appear in the Green Paper.

TABLE 2.1 Energy supply and demand, UK 1975–2000
(Assumed GDP growth = 2¾ per cent p.a. Units = million tonnes coal equivalent, primary supply)

Demand	1975	1985	2000
Energy	314	375	490
Non-energy* and bunkers	27	40	70
Total primary fuel demand	341	415	560
Supply			
Coal	126	135	170
Gas	53	65	50
Nuclear (plus hydro)	13	25	95
Oil	3	210	150
Net imports	146	−20	45
Total primary supply	341	415	510
Demand minus supply	0	0	50

* i.e. feedstocks.
SOURCE: Cmnd 7101 (1978) p. 95.

The table is expressed in terms of *primary* fuels – i.e. fuel measured in terms of the units that will be supplied. This is of course different from *heat supplied*, which will measure energy in terms of its heat content as the energy is supplied to *users*. The obvious discrepancy will be in the conversion and transmission losses between primary energy and heat supplied. Strictly, demand is best measured in *useful energy* terms, the amount of energy required for a given activity or job of work to be performed. Here again, useful energy will be less than heat supplied since there will be inefficiencies in the appliances which take the supplied heat and turn it into the heat that consumers want for a specific purpose. Thus, an insulated room will produce a larger gap between heat supplied and useful energy (assuming standards of comfort remain the same) than will an uninsulated room. Whether useful energy makes complete sense as a concept applied outside contexts such as this (essentially ones within the domestic sector) is debatable. None the less the concept, or one like it, is essential since consumers do not demand primary energy, they demand useful energy. The primary energy figures are therefore obtained by 'working backwards' from useful-energy demand projections via conversion efficiencies to the primary source implied by the demand.

THE NUCLEAR PROGRAMME

Two features of the table are of interest in the context of energy debates that are likely to occur. First, there is the apparent 'gap' between supply and demand in the year 2000. Second, there is the marked growth of nuclear power. Ancillary interests are not considered here. One would be the degree of energy conservation which is built into the forecasts, but on which there is considerable disagreement as to what is both possible and plausible. Another is the observation that oil is *exported* in the 1980s rather than conserved for later use.

What used to be called the 'energy gap' and is now called the 'policy gap' (the former phrase was abandoned when it was recognised that if the supplies were not there the demand simply could not be met, other than by imports) is seen to be 50 million tonnes coal equivalent (mtce) in the year 2000. This could easily be reduced if the forecasts are changed in a downwards direction (as some current commentators seem to think they will be: the Green Paper forecasts are already lower than those in the Energy Policy Review that preceded the Green Paper). In so far as it persists, it seems to be commonly agreed that the 'gap' would be filled by electricity. The current 'establishment' view (much disputed) is that nuclear electricity is now cheaper than coal-fired electricity (Energy Commission, 1978). It seems reasonable, therefore, to suppose that on the official policy, the gap of 50 mtce would be met by nuclear power, 'social acceptability' permitting.

We may now attempt a rough translation of these figures into actual nuclear power station construction data, since it will be argued that both the entire nuclear *programme* and its constituent parts, the single investments, are likely to be candidates for public debate nationally or locally.

The Green Paper scenario implies 41 GW of installed nuclear capacity in the year 2000. This appears to be comprised as follows:

TABLE 2.2 Nuclear capacity in the year 2000

	Existing magnox reactors	4GW
+	At completion of AGR* programme	6GW
+	Extra capacity to 2000	35GW
−	Retirement of magnox plant	4GW
=	Installed capacity in 2000	41GW

* Advanced Gas-cooled Reactor.

There is apparently some dispute as to whether the magnox retirement figure is correct or not but we have no information with which to challenge it.

The scenario as extended in Energy Commission Paper 5 (1977) suggests 135 GW (approximately) installed capacity by 2025. We assume the 'gap' of 50 mtce in 2000 (about 15 GW) is met by nuclear power. Thus we have:

$$1978\text{--}2000 \text{ new capacity} = 56 \text{ GW} \quad (41 + 15)$$
$$2000\text{--}2025 \text{ new capacity} = 79 \text{ GW}$$
$$\text{Capacity in } 2025 \qquad 135 \text{ GW}$$

To work out site acquisition we need to know whether reactors will be built on existing sites, what the size of individual reactors will be, what the generating capacity of a complete site would be, and what the replacement rate is. To make matters more difficult, it seems likely that sites may become larger in terms of GW capacity as time goes by. We therefore make an assumption that sites will be of the order of 2.5 GW minimum and 4 GW maximum to the year 2000. Thereafter they may reach the magnitude of 8 GW. We understand the technical limit is set by the waste-heat properties of large sites. On these assumptions it will be necessary to have between 14 and 22 sites by the year 2000, inclusive of *existing* sites. After 2000, an 8 GW site standard capacity would mean a further 10 sites or expansion of existing sites. If the 8 GW capacity figure is considered too large, the site acquisitions increase accordingly.

We could make a further assumption to the effect that the reactors to 2000 will, as far as possible, be built on *existing* sites. Including 'old' magnox sites and Dounreay (experimental FBR site) there are some 16 sites in all. Clearly, then, to the year 2000 it may not be necessary to acquire *any* new sites. But, more realistically, if site capacity is as low as 2.5 GW, then something like six complete 'new' sites would have to be found, or, say, one every four years. By 'new' we mean sites without existing power stations. It is common knowledge that a number of such sites are owned by the relevant electricity boards. Further, sites would not be developed to 2.5 GW straight away and new sites would be acquired early on and each would be developed to its full capacity later on.

All this suggests two scenarios to the year 2000:

(i) existing sites used to the full and expanded to about 4 GW capacity: no new site acquisition.
(ii) Some six new sites acquired with acquisitions being made early on in the 25-year period, say all in the 1980s.

Scenario (i) would suggest no major inquiries into sites, but since a 4 GW site is unprecedented in the UK (it is not perhaps the 'nuclear park' once discussed so vehemently, but it is of significant size), one might well expect some pressure for at least a local public inquiry into the expansion of sites to this size. Scenario (ii) would suggest a site acquisition every two years from 1979 to 1991. These would almost certainly become candidates for local inquiries.

In terms of *reactors*, much depends on their future size. Existing AGR stations are nominally rated at 1320 MW and, while some de-rating has taken place it seems reasonable to use a reactor size of about 660 MW (i.e. 1320 MW for a station) for whatever unit is chosen in the future. On this basis, the 56 extra GW required to the year 2000 implies at least 45 new stations, *or nearly two every year to the year 2000*. Again, this rapid rate of growth would seem to suggest that nuclear opposition groups will press for inquiries of some kind. As Chapter 3 argues, the expected rapid rate of growth in nuclear power will itself generate demands for more public participation.

There may be little point in worrrying about institutional implications beyond 2000 since, by then, societal change may require entirely new approaches. However, for the sake of the exercise, we note that, assuming what we take to be the maximum capacity for a single site (8 GW), 16 existing sites would, at that rate, supply 128 GW, which is getting close to what is required by the reference scenario. However, a number of existing sites could not take this capacity. It would seem clear that there might have to be site acquisition every three years or so from 2000 to 2025. In nuclear station terms, and allowing for plant retirement, something like 70 *new* stations would be required, *or nearly three per year*. This would be a remarkable expansion, to say the least.

AN ALTERNATIVE SCENARIO

The energy scenario in the Green Paper is but one of many. Whether it is a 'feasible' scenario depends critically on what its implications are for the rate of capital investment in energy and the degree to which this may 'divert' capital away from other areas; whether the nuclear supply industry and the coal industry can, logistically, undertake the implied rates of expansion; whether there has been any overstatement of supplies; whether there has been any understatement of demand (this is less likely – the official scenario is widely considered to be 'cautious' in that it probably overstates demand); and, of course, whether its component parts will be subject to opposition which may slow it down, whether that opposition comes from 'protest' groups, local or national public opinion, or international pressures (as with the USA initiative on reprocessing which, by virtue of the subject of our study, has not had an impact in terms of actual UK policy). While not explicitly stated as yet, the cost argument alone comes very close to an indirect argument for the early introduction of fast reactors. That is, given the possibility that energy supply growth on the scale envisaged in the Green Paper is clearly an expensive undertaking, it could readily be argued that the cost could be cut by substituting fast reactors for thermal reactors and coal-fired stations in a revised scenario. The initial capital cost per reactor would almost certainly be higher, but the use of plutonium as a fuel (and a fuel that would be readily available from the reprocessing plant at Windscale plus what is almost certainly a 'requirement' for a second one at the turn of the twenty-first century (see Pratt, 1978) would mean efficiencies 50–60 times that obtained from the use of uranium alone as a fuel. Whatever the official disclaimers, therefore, there is an argument for supposing that *actual* policy will centre on fast reactors far earlier than is suggested in official pronouncements.

But it is worth recording yet another scenario and one in which the nuclear component is not only one-third of that in the Green Paper statement for the year 2000, but which has the nuclear element actually *declining* after 2000. Moreover, the coal programme is not as high as in the official scenario for 2000 and wave power contributes a significant but by no means substantial proportion of total primary fuels. This scenario, put forward by Leach (1979) is designed to achieve an average of 2.7 per cent p.a. economic growth, consistent with the objective in the Green Paper. It claims to do so with only 70 per cent of the energy that the official scenario requires. The main achieving mechanism is conservation. Table 2.3 shows the figures for the Leach scenario and Table 2.4 compares the main elements in this scenario with those of the official one for the year 2000.

TABLE 2.3   Energy supply UK, 1976–2025: Leach scenario
(Assumed GDP growth = 3 per cent p.a. 1976–90 and 2.4 per cent 1990–2000. Units = mtce, primary supply)

Supply	1976	1990	2000	2025
Coal	119	111	122	148
Gas	59	59	55	42
Nuclear	11	36	32	28
Hydro	2	2	2	2
Oil (inc. imports)	159	171	144	100
Wave	0	0	1	11
Wind	0	0	small	6
Solar	0	1	3	7
Distributed heat	small	small	5	11
Others renewables	0	0	0	1
	350	380	364	356

SOURCE: Leach (1979). Data rounded. Wave and wind power expressed as coal equivalents in electricity generation, as with nuclear and hydro. Solar, distributed and other renewables expressed as coal equivalent for heat supply.

Tables 2.3 and 2.4 show the following interesting features: (i) on the Leach scenarios the coal industry is assumed *not* to expand more than marginally to the end of the century. Thereafter it experiences a more than 20 per cent increase in output to 2025; (ii) the gas profile is very much along the lines of the 'official' profile; (iii) nuclear power contributes only 32 mtce in the year 2000, one-third the size of its

TABLE 2.4   D of En and Leach scenarios: year 2000 (mtce)

	D of En	Leach
Coal	170	122
Gas	50	55
Nuclear	{ 95	32
Hydro	{	2
Oil (inc. imports)	195	144
Wave	{	1
Wind	{ 10?	small
Solar	{	3
Distributed heat	{ 0	5
	510–20	364

SOURCE: See Tables 2.1 and 2.3. Renewable resources are not included in the official D of En forecasts, but a 'likely' contribution of 10 mtce has been widely quoted from official sources.

contribution in the official scenario – thereafter the prospective nuclear component is actually reduced slightly in size; (iv) the total amount of oil available is significantly less than that in the official scenario in the year 2000: comparison with Table 2.1 shows that the Leach figure is consistent with the official estimate of indigenously available supply – by implication, the UK is not importing oil on a net basis in the year 2000 as it is in the official scenario; (v) very significant, the so-called 'renewable sources' appear to contribute only about 2 per cent of primary energy sources in the year 2000; (vi) the total energy needed to meet a GDP slightly less than in the official scenario is 30 per cent less, due to what, by official standards, is a formidable conservation campaign all of which Leach claims is cost-effective – i.e. yielding positive rates of return above the cost of capital.

The implications for almost any part of an energy programme are formidable if Leach is right. For our purposes, the implications for decision-making are also very significant. In essence, one could expect far less controversy over such a programme, given its standstill policy on coal to 2000 and its limited nuclear component with the promise of an actual rundown, or partial rundown, of this component after 2000. While wave power will have environmental impacts, especially in terms of conflicts of use in coastal approaches, it is unlikely to generate opposition on the scale envisaged for other energy sources. (See Appendix to this chapter for the implied nuclear construction programme in the Leach scenario.)

The major constraint on the 'plausibility' of the Leach scenario is its heavy emphasis on conservation over and above that envisaged in the 'establishment' scenarios. These doubts are not technical, or even economic in the sense that Leach argues that his programme is cost-effective (and would not be introduced otherwise). Rather they lie in the behavioural assumptions required, especially on the part of house-holders, for conservation to be introduced. Critics doubt if house-holders will insulate their homes, even with grant-in-aid schemes, to the level required. Clearly, for publicly owned housing such changes could be brought about through changes in the building regulations. Many other kinds of instrument exist for owner-occupied housing – changes in mortgage facility regulations whereby mortgages are advanced on 'better' terms for insulated houses, and so on. Just what the full range of instruments required to bring about such a scenario will be requires careful examination; and it is no part of the current work to decide between scenarios.

What matters is that, while to some the Leach scenario will not be believable, it represents a well-considered alternative scenario, thoroughly researched and at the very least deserving of wider consideration. For our purposes it demonstrates the *possibility* at least that (i) scenarios can diverge markedly even for the attainment of a given growth objective; and (ii) in so diverging they can minimise the rate of growth of investment in areas which must be considered as having the greatest potential for conflict. Such scenarios therefore have attractions on two grounds – they continue to sustain what many believe to be the ingrained social desire for growth, and they achieve it at minimum social disruption.

It may be that full appraisal of such scenarios on the same basis as that suggested for official scenarios would show them 'implausible' on one or more grounds. It may be that both types of scenario are plausible, in which case social choice should be dictated by their economic and social cost as discussed previously. It may be, and this is a real alternative that seems to lack serious consideration within official circles, that neither these scenarios nor any others designed to sustain a 3 per cent rate of economic growth are plausible. If so, one would want to look at the 'low-growth' scenarios envisaged by official sources and by the increasing number of independent and cause groups looking at alternative energy sources. Whatever the outcome, such an appraisal is surely the most fundamental part of an energy debate and one which the public should be involved in.

Appendix to Chapter 2: The Nuclear Construction Programme in the Leach Scenario

Assuming a 0.72 load factor, the GW of nuclear capacity in the Leach scenario proceeds as follows:

	Net GW capacity	=	Installed from previous period	+	New GW installed	−	GW retired
1976	4.6	=	4.6	+	0	−	0
1990	13.6	=	4.6	+	9.0	−	0
2000	12.3	=	13.6	+	2.5	−	3.8
2010	11.8	=	12.3	+	2.5	−	3.0
2025	10.6	=	11.8	+	3.8	−	5.0

Note that for an 'average' 1.3 GW station this implies some seven stations between 1976 and 1990 (the current construction programme plus AGRs at Torness and Heysham), two stations for 1990–2000, and then *one* every five years from 2000 +. Thus, while capacity declines, the 'industry' does have a limited expansion programme equal to one station every five years after the current construction phase. There are no fast reactors.

REFERENCES

Cmnd 7101 (1978)	*Energy Policy: A Consultative Document* (HMSO, London).
Energy Commission (1977)	*Energy Forecasts: A Note by the Department of Energy*, Energy Commission Paper No. 5 (Department of Energy, London).
Department of Energy (1978)	*Energy Forecasting Methodology*, Energy Paper No. 29 (HMSO, London).

Pearce, D. W.
(1978) (ed.)

L. Brookes (1972)

A. Lovins (1977)

Energy Commission
(1978)

R. Pratt (1978)

G. Leach (1979)

The Valuation of Social Cost (Allen &
Unwin, London).

'More on the Output Elasticity of Energy
Consumption', *Journal of Industrial
Economics*, November.

Soft Energy Paths (Penguin, Harmondsworth).

Coal and Nuclear Power Station Costs,
Energy Commission Paper No. 6
(Department of Energy, London).

'The Economics of Reprocessing Oxide
Fuels in the UK: a Case Study of
"THORP" ', M. Litt. thesis, University
of Aberdeen.

*A Low Energy Strategy for the United
Kingdom* (Science Reviews, London).

3 Efficient Decision Procedures

THE DEMOCRATIC REQUIREMENT

Anyone can make decisions. Deciding not to decide is to make a decision and so is choosing by sticking a pin in a map. But most decisions – acts of choosing something rather than something else – are made on the basis of information. In turn, that information may be limited or comprehensive, given the state of knowledge. Whatever its technical status, no decision can be held to be 'efficient' if it omits a major piece of information, namely, what the public wants. This type of inefficiency is a failure to honour the most elementary and fundamental feature of a democracy – that decisions should be responsive to public wants. Along with most who have attempted recently to formalise normative concepts of efficient decision procedures this underlying axiom is taken to be paramount.[1] See Arrow (1963) for a general statement, and Dorfman (1973) and Haefele (1973) as examples in the field of environmental management.

It is readily acknowledged that public wants are not easy to discern, the more so if the issue in question does not divide the main political parties, for then the usual mechanisms for articulating such wants – policy statements, debates, arguments in the mass media, etc. – tend not to operate. In the UK the debate over nuclear power has until now been an example and it is perhaps partly for this reason that the debate has emerged slowly and largely by extra-Parliamentary means, of which the WPI has been the most significant aspect. It is not argued that the WPI and the associated debates in Parliament and elsewhere served to establish public wants in this area in any definitive or permanent manner. Nevertheless, this procedure, imperfect though it is, did allow informed and articulate sections of the public to express their wants and it allowed others to express their disagreement. The resulting decision (by Parliament) could then be more efficient because public wants had, albeit imperfectly, been partially ascertained.

21

The fundamental requirement of democratic responsiveness may strike many as naïve. We live, after all, in a technological age. The public, it is argued, cannot possibly be informed of the technological basis for a given decision. Some decisions must perhaps be taken quickly. Others require an overriding 'national interest' to be used, whatever public wants appear to be. To consult the public is to engage in a certain activity of delay. On this analysis it is right that the public should be consulted only occasionally, to renew a political remit or when necessary or politically expedient. The electorate has, after all, elected a government and a Parliament to remove the onus of expressing wants directly. All this is true, at least to an extent. Yet it is no historical accident that extra-Parliamentary institutions have emerged precisely because all decisions cannot be made by a Parliament incapable of handling all issues. The public can also improve decision-making by raising issues that decision-makers may have overlooked or under-played. And if technology is complex, it is not demonstrable that the public are incapable of understanding the basic principles of technology. The simple fact is that precious little attempt is often made to inform the public of technological issues. Only in recent years in the United Kingdom have we even witnessed the release of documents which provide some of the background to important policy statements, a release formally embodied in instructions from the then Head of the Home Civil Service in July 1977.[2] This is in stark contrast to the United States, where a Freedom of Information Act obliges the Government to release documentation on request, unless it is commercially damaging or of security importance.

The doubts surrounding public information programmes seem to emanate largely from those who fear a public reaction divergent from their own opinion. Or they feel that simplification of complex issues amounts to falsification and hence to a wrong basis for the formation of public opinion. The former fear is inadmissible within the democratic framework dictated by the basic axiom stated above. The second fear is legitimate. It is difficult, however, to see that its benefits are less than its costs, especially where we have ample evidence that decisions left to civil servants and politicians alone may be just as likely to be wrong. For example, there never was a public debate on the commencement of a programme of construction of Advanced Gas Cooled Reactors, yet two thorough *ex post* studies suggest that the programme was a mistake (Henderson, 1977; Burn, 1978).

Given this basic axiom, any decision-maker may face a matrix of circumstances. The expressed public want may rest on poor technical information. In these circumstances the moral is clear – the information must be the best available given the state of knowledge. If technical expertise points to one decision and the public want to another, that conflict is properly placed before a delegated assembly such as Parliament, but the fundamental axiom stated above implies a heavy weighting of the public want *vis-à-vis* what technical experts say. Technical opinion should also be 'fed back' to the public. There can, however, be no guarantee of a convergent process in this respect. More likely is that the public will fail to express a homogeneous want – it will state differing views since wants will reflect differing *values*. Again, no decision procedure is efficient if it is not apprised of what these differing values are, and the extent to which they are held. This is all the more important in a context in which there can be no conflict resolution – i.e. where there is no compromise of views such that a middle course can be taken. The nuclear debate is characterised by this feature – the likelihood of an equilibrium of views is remote. Indeed, Breach (1978) takes the view, speaking of nuclear debates in general, 'More important and ominously, it is a conflict in which there can be no winner on present reckoning. The main protagonists can achieve their ends only at the expense of peace or freedom – or both' (p. 16). None the less, decisions must be made, although it is right to consider the phasing of those decisions. The design of institutions capable of for ever postponing decisions is formally equivalent to giving the balance of power to those who oppose whatever development is under consideration. No efficient process can tolerate such in-built bias.

Informing the public may well generate wants where none existed before. As Rowen (1976) has remarked:

> Many people who do not think much from day to day about the decline in the number of whales or black-footed ferrets, when presented with data and analyses that record and predict their extinction, may come to feel that this is a problem about which something must be done. (p. 145)

Which is the better situation – to decide on the basis of an expression of indifference based in turn on comparative ignorance, or to decide on the basis of wants generated by the creation of information? Much depends on how the information is presented, but, if for the moment we can advance the idea that it is presented 'neutrally', we would regard it as self-evident that the fundamental axiom of democracy – the responsiveness of social choice – is best met by an informed public.

It will be evident, then, that the concept of democracy underlying the preceding remarks is one that is alien to much twentieth-century practice within the United Kingdom. Unless a Freedom of Information Act akin to that in the United States or, less satisfactorily, that in Sweden, is introduced, it is clear that the public cannot be in possession of the information necessary for them to express informed wants. Nor is it evident that the Government's actions are responsive to public wants even where the latter are expressed other than at times of political sensitivity – e.g. where there are minority governments or where an election is pending. This interpretation is consistent with early interpretations of governments as vote-maximising entities (Downs, 1958). In the United Kingdom system, however, there are usually lengthy periods in which a single government may not honour the basic responsiveness axiom since taking actions counter to what the public wants is unlikely to cause a change of administration. This is especially true if Parliamentary voting is on party lines with limited free votes.

Neither, however, can the responsiveness axiom be allowed to dominate to the extent that the democratic principle underlying policy becomes purely populist. It must be accepted that what the public wants is not always in its own interest. This raises the question of the role of information again. What is essential is that the clear implications of policy alternatives are presented on the basis of the fullest possible information. We return to this point later.

In the energy context one important factor relating to the democratic axiom must be considered. While many energy decisions have a fairly immediate effect and therefore impact on the current generation, others will have their effect on future generations. This is evident when one considers the long lead-times for research and development, the long construction periods for, say, a nuclear power station and perhaps an additional period for 'proving' the technology. Two examples in the nuclear context relate to the fast reactor and to fusion power. In the United Kingdom context, the fast reactor does not exist on a commercial scale. A 250 MW station which generates electricity for consumption operates at Dounreay in Scotland. Its prime purpose, however, is as a research installation. A full 1200 MW station is planned. If we suppose that it is officially given planning permission in the early 1980s (since it is expected to be the subject of an inquiry by then), it will take at least seven years to build and perhaps another four to 'prove' itself. On the most optimistic estimate, therefore, the first reactor is not on full commercial 'line' until 1991 and a date later than that, given the delays in constructing thermal stations in the past, seems far more likely. In essence, it would appear that fast reactors belong to an energy supply system at best in the early part of the twenty-first century. Casual inspection suggests this is beyond the expected lifetimes of most of those engaged in advocating the fast reactor. The fusion process must be dated later still, given that nothing other than laboratory prototypes exist. In both cases, therefore, energy decisions are being made for a public that may exist but does not have the capacity to express a want, or for a public that does not yet exist.

Where decisions relate not to the wants of current generations but to the expected wants of future generations it is important that the decision-making process, to be efficient, should be apprised of the inherited alternatives for the future generation. It is self-evident that decisions must be made since to postpone decisions on energy installations is equivalent to saying that future generations should not inherit that energy source. The current generation therefore has no option but to act on behalf of future generations. In so doing, care must be taken to consider the costs and benefits not as seen from the standpoint of the present but as they may be seen by the affected generation. In the language of economics, it is not correct to apply a positive discount rate to such investments, or their effects, since such a procedure downgrades the wants of the future in favour of those of the present. *Ex hypothesi*, that is not what the decision is about.

It may therefore be justified to override the first axiom of democracy as we have stated it, on the grounds that, on some occasions, expressed public wants have not given consideration to the affected persons. Essentially we require that current generations act in a collective context and express a 'social' want and not the want they would express if they considered their own welfare alone. That these wants differ is fairly certain and forms the foundation of a number of social choice theories (Sen, 1967; Harsanyi, 1953; Marglin, 1967). One possible source of guarantee that they will so act is that the next generation's welfare is likely to be a component part of the current generation's welfare. In the economist's jargon, utility functions for the current generation 'contain' the utilities of future generations as arguments. At the level of simple language, our children's welfare is part of our own welfare. How far this interdependence of wants through time can be relied upon *alone* is debatable. If it can be relied upon, the fundamental axiom of democracy is not altered. Where the judgement is that the public want is not coincident with the collective choice, then the elected body has the responsibility of interpreting collective choice.

This phenomenon whereby the public want is superseded by some central judgement of what the collective interest should be is unlikely to be dominant in energy decisions. However, one feature that may result as a consequence of energy investments is potentially important. Special attention needs to be paid to the phenomenon of *reversibility* (Krutilla and Fisher, 1975). Most investments will have benefits and costs which are reversible. If a future generation does not like a motorway, it can, technically, cease its use, possibly dig up and reclaim the land for, say, agriculture. In the same way, nuclear power stations or coal-fired stations may be closed down. It is less obvious that a future generation inheriting vitrified radioactive waste can reverse that inheritance if it so wishes. If cumulated CO_2 in the upper atmosphere caused by fossil fuel conversion is a problem in terms of its climatic effects, then it is also not clear how a future generation can reverse these effects.

Does this mean that decisions involving irreversibilities should never be taken? This would be one extreme view, sometimes encompassed by the advocates of what has been known as 'open-ended planning' in which no reduction in the choices of future generations is permitted. That is, any choice made now must be made in such a way that it has the certainty of an associated negating choice attached to it: a later generation, or the same generation at a later date, can reverse the choice and return to the original situation. In reality, such a rule of choice would make decision-making impossible. No destruction of unique landscape would ever be permitted since no one can restore an original unique natural asset. Of course, some form of near substitute may be achieved, as with the restoration of coal tips and so on. But the rule would, strictly interpreted, entail that complete allowance for 'regret' be made in a decision-making. Anything that is regarded as a benefit now may be regarded as a cost in the future. Conceivably, the reverse would be true in that what is regarded as a cost now should be thought of as something which may be regarded as a benefit in the future. Planning to allow for this type of reversibility is likely to be so restrictive that few decisions would be made. Of course, this may be the attraction to some of such a planning paradigm since it would effectively preserve the *status quo* and could certainly be construed as an anti-economic growth standpoint.

More realistically, it makes sense to consider future costs in the light of the reversibility requirement. Algorithms for investment choice when some consequences are irreversible have been provided by Krutilla and Fisher and others at Resources for the Future, Washington. It must be made clear that the presence of irreversibility does not, on their analysis, preclude an investment decision being in favour of development and against preservation. What it does do is to impose somewhat more severe restrictions on development where irreversibilities exist.[3] What is being suggested here is that decision-making procedures are not efficient if they fail to incorporate some mechanism which reflects the extra penalties that should be placed on investments with irreversible features. That mechanism need not be a formal investment rule, as in the Krutilla/Fisher case, although it could be argued that the application of such rules should be made in any cost-benefit study or impact assessment.

We may now sum up on the first and major requirement for an efficient decision process. This requirement is that the procedure should be reponsive to public wants, with the proviso that some judgement be made that expressed public wants allow for the collective features of some investments, especially in so far as they affect future generations. This statement effectively encompasses most people's idea of what democratic procedures would imply in this connection. It is important, however, to recognise that it does not accord in full with democratic institutions as they now exist in the United Kingdom. Here a permanent bureaucracy is very much at the heart of decision-making and the 'appeal to the public' is an irregular and unsystematic feature of the process. This effective 'closure' of government is merely reinforced by the non-disclosure of information on how decisions are made, albeit with some indications of a change in this aspect of government. If these procedures are followed they would lead to greater public participation. That in turn entails that the public be informed; and we turn to this issue in more detail later.

The search for improved democratic institutions may be considered by many to be an idle pastime. They might argue that the system has evolved in an adaptive way in the United Kingdom to give us the best possible, albeit still imperfect, system. The search for better institutions to meet some set of criteria for 'efficiency' could therefore be fanciful. One other concern is that demands for institutional change are demands for the creation of a system which can only come up with the results that the advocate himself wants. In response it could be suggested that failure to change institutions reflects a desire to maintain a system that is designed to produce the opposite answer – i.e. to confirm the 'establishment' in its existing policies. There can be little doubt that there are advocates of alternative energy futures and, indeed, alternative futures full stop, who do not recognise the basic laws of argument or the basic axiom stated at the outset of this chapter. There can also be little doubt that there are advocates of 'official' energy futures who would resent wider participation in policy formation precisely because it might conflict with their own predispositions. That neither group should be permitted to dictate policy should be reasonably self-evident. Proposals for change should be seen in this light.

It is worth stating ideals towards which the system might work. It would perhaps be presumptuous in the eyes of many to suggest wholesale change. Since that is not what is being suggested here, this charge is not one that has to be faced. Finally, the concept of adaptive evolution of institutions ignores the very process by which change is brought about. It is not an exogenous, 'natural' force, but rather one that responds to calls for change. It is all too often forgotten that many current debates only occur because of the requests and pressures of those who seek change. And institutional change has occurred also on exactly the same basis. While it may therefore seem to some a grandiose activity, the contemplation of institutional change for more efficient decision-making should perhaps be a wider activity. Within the energy debate it is significant that Government announcements have stated that 'it will be important to reach a reasonable degree of public consensus on the acceptability of nuclear power' (Cmnd 7101, para. 10.17). That nuclear power has been accorded 'special treatment' by the Government is a reflection of the political importance attached to nuclear power issues. That importance may be much exaggerated in comparison to other environmental issues such as metal poisoning (Schroeder, 1974; Nobbs and Pearce, 1975), but it has none the less arrived in the UK as a national political issue, and merits serious study.

A MODEL OF AGENDA SETTING

Easton (1965) has proposed a model which formalises the various levels at which expressions of individual want become translated into items which appear on a political 'agenda' – i.e. become important enough to attract special attention from central or local politicians or, in our case, important enough to merit some revised decision-making procedure. It is useful briefly to outline Easton's categories. A *want* in his language is simply some feeling for change, a concern that the desired state of affairs diverges from the existing state. Wants are clearly related to complex factors such as socio-economic status, culture and inherited or absorbed values. Wants only become *demands* when the want is voiced as a requirement or request that the existing state of affairs be changed so as to bring about a new state. Demands become *issues* when they are taken up by collections of individuals or groups. Nuclear power thus becomes an issue because it is the source of concern of groups who adopt a 'cause' centred on the issue in question. Issues only become items on the *agenda* if they are formally accepted as being deserving, for whatever reason, of public debate and political action.

It is important at this stage to distinguish two types of 'agenda'. Taking the example of nuclear power, it has always been a matter of national political concern if only because of its intricate links to military nuclear power. But that political concern has been expressed within government rather than by the public at large, even where the 'public' can be identified with professional 'cause' groups which associate nuclear power with environmental or political risks (radiation, proliferation of nuclear weapons, the centralisation of power and so on: for an analysis of the types of strengths of the concerns see Otway, 1978). In the same way, coal policy in the United Kingdom has always been on the national agenda in so far as there has been a planned reduction in the size of the industry and now a planned expansion of it. In both cases the issues have been debated within the civil service, within government and within Parliament. Indeed, it is significant that in the UK nuclear power was seen as something meriting centralised control from the start to the point where free market principles have intruded rarely, even in the domain of such issues as reactor choice. For some, e.g. Burn (1978), this is a matter of regret and the industry as a whole would, he argues, have been all the more efficient if the industrial structure and control of R&D were not as centralised as they are. Whatever the truth of this contention, it is evident that energy sources in general have always attracted centralised political attention.

What has *not* reached the agenda until recently has been the wider issues which we might loosely call 'environmental' and 'socio-political' risk. Here the impetus has come from outside government and the bodies controlling the energy source or involved in supplying its capital structure. It has come from the environmentalist movement. Thus, what concerns us here when we speak of matters getting 'on to the agenda' are the twofold aspects of (*a*) debating the entire social implications of an energy source or entire energy programme, and (*b*) the *province* of that debate being extended to a public which had hitherto not been involved in the politics of energy. It is as well to bear this important distinction in mind in what follows.

Clearly, nuclear power is an issue that has got on to the political agenda in the sense used here. Yet other energy issues have not, or at least command far less time on the agenda. It is not evident, for example, that coal policy has anything like the same political 'status' as nuclear power policy. Few groups espouse coal as an energy source of concern in the sense that they wish to vote against its expanded exploitation. Some espouse coal as a *desirable* energy source, frequently, one suspects, because they cannot be seen to be against all 'conventional' sources of energy and because one extra coal-fired station might, to them, be one less nuclear power station. Others would point to the prospect of technological change in coal-fired power generation that would remove most of the pollution aspects – e.g. fluidised bed combustion.

Yet, if a comparison is made between coal and nuclear power it is not immediately evident why one should be 'on the agenda' and the other not, or less so anyway. In terms of pollution, landscape despoliation is greater with coal than with nuclear power. Immediate air pollution appears as a debit to both types of power station in the sense of pollutants that affect the local community. Spatially distant air pollution is possible in both cases, with radiation hazards showing up in the food chain and SO_2 from tall chimneys in the form of acid rain in Norway. CO_2, could, on some analyses, constitute a major long-term pollutant, while radioactive wastes vary in their half-lives from hours to nearly one quarter of a million years. Both have irreversible aspects if the CO_2 argument is correct. If it is not correct, then the reversibility criterion favours coal more than nuclear. In terms of workforce injury a proper comparison is with uranium miners and coalminers, coal distribution and conversion and the nuclear fuel cycle beyond the mining stage. It is not evident that coal would fare better in such a comparison.

Two questions can therefore be asked. *Should* one energy source be 'on the agenda' and the other not? If the answer is 'no' then the options are that *both* should be on the agenda or that neither should. Second, why is it that one *is* on the agenda and the other is not? Again, bear in mind that this use of the term 'agenda' embraces an *additional* set of debates to those which have historically been the subject of political concern, *and* an additional number of participants.

There is a serious reason why nuclear power should appear on the agenda regardless of whether coal does or not. Nuclear power is not readily divorceable from the military use of its fuels. The whole issue of nuclear proliferation via the use of fuels 'diverted' from civil use is a complex one (we discuss it in a little more detail in Chapter 7). None the less, it cannot be denied that a problem exists. Escape from this problem via 'technical fixes' such as the irradiation of fuels for return to countries of origin are therefore of importance in considering just what the scale of the problem may be. For the moment, it is only necessary to note that the problem cannot be said to be non-existent.

Simply because of the proliferation problem, including the use of civil fuels for terrorist purposes, the working population in a nuclear industry must be subject to restraints on their liberties. No one doubts that this is the case. There is therefore some loss of freedom, not necessarily confined to the workforce alone, that ensues with a nuclear future. If it were possible to substitute other fuels, it is less obvious that this loss of freedom would occur.

If therefore we add these two special features, both emanating from the relationship between civil and military use of nuclear fuels, to the problem listed as being common to coal and nuclear power we see that there is an 'added dimension' which justifies placing nuclear power on the agenda. Should coal be on the agenda too? Given the death rate within coalmines and given the likely damage done by coal conversion it should. An agenda may be national or local. In this work we are concerned with both. Local issues can be dealt with adequately by suitably modified existing institutions provided the informational requirements set out later are met. To illustrate how a national agenda might be limited consider the case of liquified natural gas. LNG accidents have the potential for being dramatic (and have been) in terms of the numbers of persons who could be injured or killed in a single accident. Safety standards should be uniform and strict and the setting of those standards would be an item on a *national* agenda. But, whether or not a local community accepts a gas shipment terminal is arguably not an item for the national agenda. *Provided* that the local population is fully informed, the damage done by an accident is something for them to appraise since others appear not to be involved. Of course, the transhipment of LNG by rail or road could be regarded in a different light. No single community is affected – many are. Moreover, a community may be completely unaware of such transport taking place within its local boundaries. Public inquiries into whether particular items should or should not be routed through a given locality are conspicuously absent. The transport of LNG could very well be an item for the national agenda. For the local LNG plant, however, the damage seems to be local and has no real national dimension.

Against this, one might put the following argument. If national standards are laid down for an LNG terminal and if a local community 'accepts' such a terminal, then would the resulting accident, if it occurred, be a purely local issue if it generated widespread concern among other localities? In the first place it might suggest that safety standards, agreed to be an item for the national agenda, were not adequate, or that the risk-benefit trade-off was not nationally acceptable in so far as any of a number of communities might face the siting of such terminals in the future. Second, and a development of the first point, perception of risk is altered noticeably when the risk actually materialises.

Consequently, there might be a more than local demand to site such terminals away from centres of population in exactly the same way that nuclear power stations have traditionally been sited in the UK, in general. The issue is obviously complex. The difference between setting the *rules* for siting LNG terminals still needs to be differentiated from the acceptance or rejection of a specific installation by a single community. The overlap remains however, since clearly that process of local rejection or acceptance may be critically dependent on the perception of the acceptability of the risks implicit in the national rules. Moreover, posing the issue in the way we have could also be deemed misleading in that we have presumed a single site application by, let us say, some oil company concerned with finding a site for the terminal. Rather the issue might be turned round to suggest that it requires inquiry procedure which is *not* constrained to consider a single site but many, in order that the risk-benefit trade-off be optimised. Just such a planning inquiry procedure exists in the UK, on the statute books anyway, and this is the Planning Inquiry Commission. Quite why it has never been invoked in practice is an issue we address ourselves to later. To summarise, it may well be that the siting of LNG *plants* is a matter for the national agenda, even though the risks are ultimately borne locally and the benefits nationally.

More formally, issues that deserve to be placed on the national agenda are those that have national dimensions which can comprise one or more of:

   (i) investments which have spatial environmental consequences beyond those of a local jurisdiction;

  (ii) investments which have environmental consequences beyond a time horizon which lies wholly within one encompassed by the current generation;

 (iii) investments which have implications for traditional or desirable freedoms of the individual;

 (iv) investments otherwise generating national concern in respect of the rules and guidelines for siting them.

In the economist's language, public goods and 'bads' belong to local decision-making procedures and therefore local agendas if the external effects associated with those goods and bads are themselves localised. Public goods which have national or even global externalities attached to them belong on the national agenda. (Note that we are not concerned here with the definition of the optimal jurisdiction for a public good – the rules above are general ones and relate to *existing* structures of local and national government. We have therefore taken the governmental structure as given and have worked on the basis of which 'goods' should be considered within which jurisdiction. We have not considered taking the goods as a datum and changing the jurisdiction to deal with them.)

Two aspects are worth noting in respect of the above discussion. First, whether an issue should be on the national agenda is determined by its perceived national benefits and its costs. Dispute is however more likely over the costs than over the benefits, although in the energy debate there is one very important issue relating to benefits. The issue will be whether the perceived benefits – let us say, some given increment in GNP brought about by having nuclear power – outweigh the costs in the sense of both the resource costs of the investment and the associated external costs. There will be technical disputes about the costs, especially where they are probabilistic, as is inevitably the case with nuclear power (e.g. the risks of cancers, or the risks of war caused by a state that acquired a weapons capability from the use of civil nuclear fuel). There will be differences in the evaluation of the costs – e.g. differences in importance values attached to any infringements on civil liberties. And so on.

Benefits will be disputed in a somewhat different sense. Here the dispute is not likely to be over the size of benefit – although this will be the case for such things as employment impacts when comparing nuclear power to, say, alternative energy sources. It will, rather, be over the issue of whether what some people regard as a benefit is regarded as a benefit by others. The most obvious example, although its relevance to nuclear debates still appears to have been misunderstood by many, is that a claim that nuclear power will enable a given level of GNP to be sustained will be seen by some as a *cost* of nuclear power. This is simply because they object to the pursuit of given positive rates of change in GNP as a policy objective, arguing that growth induces values which are inimical to a 'desirable' society, or a society with a long survival time. This is important. The dispute over costs is likely to be a dispute over quantities and to some extent values. The disputes over benefits is much more likely to be a dispute over values and less so over quantities.

The second point to be noted is that, in terms of the formal model used, many wants will not become demands and many demands will not become issues. In both cases, the wants and issues will not therefore even appear as candidates for the national agenda. This remark is necessary because the most common objection to justifications for making a national issue out of a given policy or investment is that it exaggerates its importance and gives rise to the potential for never-ending debates which effectively preserve the *status quo*. If, say, it could be argued that nuclear power was not deserving of national agenda status it would be possible to proceed more rapidly with investments and thus secure the benefits earlier. If the advocates of the policy are correct in their assessments of costs, this will raise the expected value of the net benefits (increase the discounted present value) if a discount rate is used. If the discount rate is zero, then of course, delays will not affect the present value unless they reverberate through the economy in some manner which is not invariant with time. Our argument is that issues such as nuclear power *should* be on the national agenda and this argument is unaffected by the fact that nuclear power *is* an issue. That what *is* need not influence what *should be* on this model is illustrated by the fact that we regard coal investment as being a candidate for the national agenda (in the sense used here) when in fact it generally is not in the United Kingdom. That is to say, we believe that if the problems associated with coal exploitation were more widely appreciated coal exploitation would *become* an issue and would be on the national agenda. This relates to what we have to say about informing the public (see later). But, by suggesting changes in the institutional structure of decision-making procedures, or rather, in the advisory procedures that lead up to the decision-taking level, it must be fully recognised that one will generate more issues than would otherwise be the case. This is not, however, a defect of the proposed institutional change. Rather it is a benefit, simply because it does bring into the democratic process policies that would otherwise remain within the structure of bureaucratic decision-making, and therefore largely unaccountable to the public. That information may generate wants, demands and issues is not a cost: it is entirely within keeping of the democratic axiom enunciated in the early part of this chapter.

The second question asked of agendas was why some issues rather than others actually *do* get onto agendas. One answer can now be that some issues get on the agenda because they merit attention at that level. Certainly nuclear power deserves such attention on the basis of the reasoning outlined earlier. None the less, this hardly explains the whole sociological phenomenon of there being a movement that is 'anti-nuclear' together with the decided absence of a movement that is 'anti-coal', if it is agreed that both should be on the agenda. We offer little by way of explanation here. We doubt if anyone has come up with a satisfactory theory, although countless explanations are offered: notable among these is the hypothesis of guilt-ridden societies looking for something on which to assuage their guilt – exemplified by shame over the Vietnam war and an urge to rectify the situation by engaging in some activity which requires the demonstration of a social conscience. Even if this is true it hardly accounts for the international nature of nuclear opposition. Nor would it be clear exactly what guilt is being expunged in the UK anti-nuclear movement. Opposition could just be a mistake in the sense that it may rest on ignorance of the safety of nuclear power, an ignorance fanned by the association of the atom as a source of power and the atom as a source of a devastating weapon of war. Such ignorance may also have been expanded by the secrecy surrounding nuclear power until recently. The hypothesis is plausible, but scarcely accurate in its assessment of some of the opposition which is articulate and intelligent. It may be that issues *become* issues for reasons which are later discarded, in favour of other more substantial reasons. Perhaps affluent societies can 'afford' the luxury of opposition to energy sources that are thought to be damaging to the environment. Even though there is, contrary to the opinion of many, little to substantiate the view that the environment is an 'elite' good (Pearce 1979), it cannot be questioned that greater education and leisure time permit of a middle class that does question established policy. If this view is correct, it still leaves unanswered the problem of why this group of persons should 'pick on' one energy source and not another for its opposition.

More amenable explanations are at hand, however. While a much rehearsed argument, it is true that exponential growth entails bigger and bigger absolute levels of investment to sustain that growth. If energy and GNP are linked as closely (or even anywhere near as closely) as many commentators suggest (Darmstadter, 1978), then energy investment must increase rapidly to meet the demands of a growth society. Such a phenomenon will have various effects. It will cause a depreciation in the natural environment and hence raise the levels of awareness of those who have been used to a greater supply of such goods in the past. Quite simply, as any elementary economics textbook points out, the value of a marginal unit of something is higher the less there is of it. Such changes in awareness are disseminated because informational networks are improved, so that, in terms of our model, demands become issues far more quickly. Lastly, in an exponential world it is necessary to advance the process of technological evaluation and control at least as fast as the rate of change in the technology itself. Failure to do this entails a failure to control the process of change, the ramifications of which tend to be synergistic and exponential. There can be little doubt that one of the central features of opposition to much change is not to the change itself but to the *speed* at which it occurs. Things happen too fast for societal control, or at least too fast for society to *feel* it is in control. It is no accident, then, that opposition to nuclear power occurs just at a time when the growth of nuclear power, in terms of intent anyway, is faster than ever because of the 1973 oil price rise and the push this crisis gave to investigations into just how long conventional energy sources would last.

It may be, then, that one need proceed no further than the previous argument to 'explain' nuclear opposition, although we doubt it. Certainly, tracing the argument through for coal would not give rise to the same conclusion. Coal technology has been with us for many years, the industry is established and is 'part of a lifestyle'. While doubling coal output would be a major task over, say, 25 or 50 years, few would treat it as being a source of deep concern (although it should be). All this said, the way in which demands become issues and issues get on to the agenda remains a much under-researched area. We make no claims to having a definitive theory of how nuclear power arrived on the agenda, nor do we need one for our purposes. We do need to establish that it deserves to be on the agenda and that we would argue we have done.

INVESTMENTS VERSUS PROGRAMMES

A critical issue relates to what exactly it is that should be placed on a national agenda. Within the field of nuclear investment in the UK, one might contrast, say, one extra thermal power station in an established programme of such stations and, let us say, the first commercial-sized fast reactor in the UK, CDFR (Commercial Demonstration Fast Reactor). The former cannot be construed as a new *programme*. It does not represent a change of policy but rather a component part of an existing policy that has been under way for some years. If therefore the siting of thermal reactors is considered to be a local problem, it seems that an efficient decision advisory procedure would be fulfilled by establishing a local public inquiry. Although the benefits of one more power station (of whatever kind) are national in the sense that the station feeds a national grid system, the costs are both national and local. They are national in the sense that the taxpayer meets the bill for constructing the station, but local in the sense that the local population is most at risk in the event of accidents and in that any landscape effects are localised. Moreover, there are local benefits in terms of employment. The fact of national costs can hardly be cited as a reason for putting an extra power station on the national agenda, since this would apply to all public investment. The extra power station therefore appears to be an example of a localised good which, if it becomes an issue, is best debated within the local context. Some of the arguments discussed in the hypothetical case of the LNG terminal are relevant here, however. None the less it seems reasonable to proceed on this basis.

It is worth considering some of the problems in adopting this approach to single investments since they help to determine what issues should be placed on a national agenda. First, the 'need' for the investment is a legitimate ground for concern. Need is a national issue and, indeed, the very word implies something more than is actually the case. It implies necessity when what is really intended is a phrase such as 'socially desirable'. That is, the investment will be justified by its advocates in terms of statements to the effect that if the public at large wants extra energy then this extra power station is required. Traditionally, issues of *national* 'need', or what we prefer to call social desirability, have been excluded from local public inquiries, the emphasis being on the implications for the locality alone. The basic reason for this is that any such questioning in a local inquiry context is considered to be a challenge to the Minister's decision-making power in the wrong context. That context is held to be a national one and must ultimately come through Parliament.

More recently there has been a slight change of policy in respect of inquiries into road construction plans. Under new proposals (Cmnd 7133, 1978) the relevant Government Department will explain to inquiry participants how the proposed road fits into national road policy, i.e. why it is, in that sense, 'needed'. Further, objectors may submit statements to the inquiry Inspector (Reporter in Scotland) before the inquiry, questioning the 'need' for the road. They may also propose alternative routes, a significant change in policy.

As far as questioning the desirability of an investment it is unclear what function such submissions will serve. It seems clear that Inspectors will not be permitted to make recommendations that *no* road is needed. This is perhaps as it should be since national transport policy is something belonging to the national agenda and ought not, on the establishment argument, to be determined by an aggregation of local decisions. In the view of others this is perhaps exactly how national policy should be determined. None the less, one can see that unless all localities have inquiries (most road construction takes place without any inquiry) and unless they all vote one way – either yes or no to the road in their locality – the prospect of a road system made up of small segments of differently graded roads becomes a real one. It seems right therefore to accept the view that national road policy is something that everyone should be informed of but, if they oppose it, they should do so at the level of the national agenda. The local inquiry remains the place for a discussion of local issues. However, this much said, apart from informing objectors of the role of a given road in a national system, it is quite unclear what purpose is served by inviting or encouraging objectors to question that policy in the context of a local inquiry.

Proposing alternative routes is a different issue. What emerges here is the potential for setting one locality against another as each tries to 'shift' the burden of the road on to some other area. This process is unlikely to be too unwieldly, however, since there is a presumption that, in the majority of cases, alternative route proposals will not be accepted. Hence the process is convergent on a finite number of inquiries.

Now relate this change in policy on road inquiries to energy inquiries. A valuable informational function would be served if each inquiry was accompanied by a statement of why the new power station is regarded as socially desirable. Given the time-lags involved, this can only be done against the background of energy forecasts and both the forecasts and the methodology underlying them should be public (both now are in the UK with the publication of the forecasting methodology in 1978– see Department of Energy, 1978). It remains the case that any challenge to the forecasts and the methodology belongs to the national agenda and not to the local inquiry. We consider later whether national institutions are in themselves 'efficient' for this purpose. Should objectors to power stations be free to propose alternative sites? Having accepted the principle for roads, it is not altogether clear why it should not be accepted for power stations or even reprocessing plants. Again, the technical and economic limitations for siting energy installations should result in a convergent process, although one sees scope for considerable delaying tactics by objectors if this is permitted.[4]

Overall, then, the first problem, the questioning of 'need' ought not to increase problems for local inquiries which, we suggest, are the proper institutional context for single energy investments which do not have a national dimension. The second problem relates to a familiar argument that, if no single investment warrants being placed on the national agenda (with exceptions which we note shortly), then there is a risk that the sum of all the investments, a programme, will be secured without public debate. That is, the public would be denied the chance to debate the social desirability of roads *per se* or nuclear power *per se*, and the programme will have emerged by a process of 'incremental gradualism'. The answer here is that national institutions should be such that they ensure that any programme is debated if there is a public desire to have it debated. The proper agencies here would be Parliament and extra-Parliamentary institutions that have national functions. In the energy context these include the Royal Commission on Environmental Pollution, the Energy Commission, the Commission on Energy and the Environment, and the Select Committee on Science and Technology (House of Commons). The structure of these bodies is described in Chapters 4 and 9.

Historically, it is difficult to find a major assessment of the thermal reactor programme. That it was not debated publicly is consistent with several hypotheses. Perhaps the public were indifferent to it. Perhaps institutions did not exist to deal with it (only the Select Committee existed out of the above list of institutions). Parliamentary acquiescence could be construed as assent to the programme, and even as a reflection of public indifference if one accepts a more populist view of UK democracy than we have suggested. While information was released about the programme there was a widely held view that nuclear power was not dangerous in any respect – i.e. was riskless. Belief that the government knew what it was doing was similarly pervasive, with some exceptions. None the less, what information there was contained little on the risks of nuclear programmes. Apparent indifference may therefore have been due to lack of information, although on a simplistic model of the democratic process one would expect the public to demand the information. What is being suggested here is that 'wants' and 'demands' had not formed because of public trust in technocrats and government. In the event, it was only with the celebrated 6th Report of the Royal Commission on Environmental Pollution (Cmnd 6618, 1976) that nuclear power was looked at in any great detail. Arguably, this was enough. It seems preferable, however, to accept that the thermal reactor programme should be subject to periodic scrutiny by Parliament and by the outside institutions which now exist or which may exist in a modified system. What is true of the reactor system should also be true of national plans for coal, and other energy sources.

Such a suggestion is not out of keeping with what has already been agreed for road investment. A new Standing Advisory Committee will monitor the methodologies used for road appraisal (forecasting models, cost-benefit studies, etc.). What is proposed here for energy is entirely consistent with what has been agreed for roads. But, in addition to looking at the methodologies, the relevant agency would also look at the programme, its time-phasing and its desirability. Most of these aspects would in fact follow on from assessments of methodologies. If forecasts are wrong or are revised then there are obvious implications for programmes.

Overall, then, while single investments belong properly to local inquiries, the national policies which dictate their existence should be under regular scrutiny. Given the demands made on parliament it seems proper to suggest that some extra-Parliamentary body should have the advisory and investigatory role in this context. This in no way precludes Parliament or Select Committees from making their own assessments and of course implies nothing that reduces the power of Parliament as the ultimate decision-making body.

Now consider the second investment, a commercial-scale fast reactor. The position of the UK Government has already been made public. The substance of their view is contained in their response to the 6th Report of the Royal Commission on Environmental Pollution. There they stated:

The Royal Commission recommended . . . that a special procedure for public consultation should be established in respect of *major* questions of *nuclear development* to enable decisions to 'take place by explicit political process'. In the case of *specific projects* there are provisions for public inquiries to be held, as in the case of the proposal to build an oxide reprocessing plant at Windscale. The Government fully accept however the need for a proper framework for wider public debate, and will consider the most suitable kind of *special procedure* to achieve this, bearing in mind the Commission's suggestions in this respect. They also accept that, *before any decision is taken on CFR 1, this procedure should be settled and announced.* (Cmnd 6820, 1977, para. 40: our own emphasis)

The reference to 'CFR 1' is to what was later called CDFR. The language of the White Paper is at least clear in promising a 'special procedure' relating to a commitment to a large nuclear programme that includes fast reactors. Close inspection of the text shows that it is ambiguous on whether this 'special procedure' would apply to an additional programme of thermal reactors alone, and ambiguous on whether CDFR is also to be subject to a 'special procedure' although that procedure is to be established before a CDFR inquiry. One reading would suggest that the debate on a combined thermal/fast reactor *programme* should take place first and only after that would CDFR be considered, perhaps by a normal public inquiry process. Another reading suggests that CDFR itself will be the subject of a special procedure. In either case there would have to be two such inquiries, one into the single investment and one into the programme of which it may form part. Yet another interpretation is that CDFR and the nuclear programme would be discussed simultaneously.

Further, the 1978 Green Paper *Energy Policy* states:

> Any decision to build the first commercial fast breeder reactor would be subject to a public inquiry. The precise procedure which would be adopted for CFR 1 has not been settled *but the inquiry would not be limited to local planning matters and would enable wider relevant issues to be considered.* (Our emphasis: para. 10.6)

Again, the statement is consistent with any of the alternatives above. In actual fact a special procedure for CDFR is to be used – see Chapter 11.

It was argued earlier that any thermal programme should be the subject of an investigatory process outside Parliament. Regardless of how the Government's promises are construed, is there a case for placing a *single* fast reactor on the national agenda? The arguments against doing so are very much the same as those advanced in favour of debating single investments of any other kind at local inquiries if the issue is forced. The arguments for differentiating it from single investments essentially turn on how the first fast reactor of commercial size will be viewed by the public and interest groups. In essence, the reasons for elevating CDFR to the national agenda may be as much political as they are logically persuasive.

Thus, anyone familiar with the changing terminology used to describe the first fast reactor will be aware of rapidly shifting changes of emphasis as the Government recognition grew that early pronouncements did imply a commitment to a fast reactor *programme* whereas the heavily influential 6th Report of the Royal Commission had called for a special debate and public involvement before such a commitment was made. The history of this symbolism is worthy of a study in itself. First, the term 'breeder' was dropped from the description 'fast breeder reactor' since it was felt to be emotive and perhaps suggestive of some uncontrollable force. CFR 1 (Commercial Fast Reactor 1) was designed to indicate that what was under consideration was a commercial-*scale* (1200 MW) reactor. Since it would be the first such reactor, the numbering was self-evident. Later it seems that someone thought that numbering implied a programme: 1 implied there could well be a number 2, 2 implied 3 and so on. Since this was inconsistent with the language of the White Paper, which implied non-commitment to a *programme*, the numbering was dropped. Later still, a 'D' for 'Demonstration' was added to make CDFR, which is how it stands at the time of writing. To add to the confusion, the 'C' does not imply that the first fast reactor will be an economic undertaking: it merely refers to commercial scale. Nor is the 'D' to demonstrate that it is technologically sound but rather to demonstrate to eventual purchasers (the electricity boards) that fast reactors are sound investments. A cynic might be forgiven for asking what it is that is being demonstrated if the plant is not expected to be economic and if there is already certitude on the technological side. Arguably, the eagerness to avoid unfortunate public images led to a semantic misfortune.

Whatever one's view of this state of affairs, it is clear that in 1977 policy pronouncements switched quite dramatically from talk of a fast reactor programme to talk of a 'one-off' investment. That the change even caught out those who write policy documents is evidenced by the fact that the White Paper of 1977 continued to talk of CFR 1 whilst simultaneously agreeing to the Royal Commission's call for a debate before commitment to any programme. One reason for placing CDFR on the national agenda, therefore, is a logistic one: in the eye of the public this change of policy was so poorly handled that few will believe that what is under consideration is just *one* fast reactor.

Second, since it is the first such reactor (apart from the 250 MW Dounreay reactor) it could be argued that it is symbolic of a change in energy policy within the nuclear component of UK energy policy. In Austria it was decided to hold a national referendum on the first *thermal* reactor. By comparison, the idea of a national debate in, say, a planning inquiry commission plus an evaluation of nuclear futures by some other body is a mild proposal.

Third, much effort has gone into justifying the fast reactor proposal in terms of uranium supply constraints. Here again the emphasis has shifted somewhat. Talk of certain uranium shortage in the early twenty-first century has given way to talk of 'uncertainty' about supplies and the fact that the UK must necessarily be an importer of uranium, perhaps from politically unstable countries or countries exercising controls over uranium exports (as with Australia, Canada and the USA). But whether the talk is of definite worldwide shortages, probable shortages, or artificially created shortages through political manipulation of supplies, it is difficult to escape the impression that public pronouncements belie actual policy. Again, this suggests that CDFR is not likely to be seen as one investment divorced from the concept of a programme.

Fourth, and a very arguable viewpoint, many participants to the Windscale Inquiry in 1977 saw, and continue to see, THORP as a plant designed not just to deal with the waste disposal problems of thermal oxide fuel reactors, but also to inventory a most valuable fuel, plutonium, again on the probabilistic argument that it will be 'needed' in the twenty-first century. It matters little if the first few fast reactors can be fuelled by stored plutonium already in existence. The issue is simply that THORP can perhaps be made to appear economically a most attractive investment if part of its output is a fuel which has enormous potential as an energy source. At any reasonable discount rate and on any reasonable probabilities about uranium supply and price, plutonium storage *may* make economic sense. If this argument is even remotely correct, CDFR is again symbolic of a programme, not a one-off investment. Note that a House of Commons Select Committee has itself stated: 'It [CDFR] would represent the first step towards the domination by fast reactors of the national energy supply strategy' (Select Committee, 1978).

For these reasons alone CDFR should merit a place on the national agenda. It must be accepted that the reasoning is slightly unusual. In the first place it suggests that beliefs have a habit of changing more slowly than public pronouncements would wish. This seems true in this case and is in no small part due to the rapid and slightly confused turnround in policy in 1977.[5] Second, it may seem to suggest that what is actually going on is a deliberate 'cover-up', an attempt to mislead the public. This is not the suggestion here. It is evident that those responsible for policy have attempted to respond genuinely to a persuasive call for a wider public debate made in the Royal Commission's 6th Report while simultaneously having to alter public statements to reflect that response. It would be surprising if such a rapid change could have been made without leaving the anomalies noted above, since these arise precisely because they are the heritage of prior policy which implicitly accepted fast reactors as inevitable.

Finally, while singling out one investment for placement on the national agenda, the reasoning above is not likely to apply elsewhere.

Some people have suggested, for example, that exploitation of coal in Belvoir and Selby should be debated as a national issue. In terms of the analysis here, these investments as single projects would be the subject of local inquiries and the arguments put forward for placing a fast reactor on the national agenda would not apply to them. Equally, however, the issue of 'need' determines whether these investments should be undertaken. At some point in the decision-making process therefore, the need, or social desirability, of expanded coal investment must be discussed as an item on the national agenda. The same remarks therefore apply to a coal programme as to a thermal reactor programme or a programme for LNG and so on. (The Commission for Energy and Environment commenced a programme of work on coal policy in 1978.)

INFORMATION AND PARTICIPATION

The outset of this chapter stated a fundamental axiom of democracy, that public decisions be generally responsive to public opinion. In turn, we argued that public opinion, to be conducive to efficient decision-making, must be informed. We have seen acceptance of this principle, in part at least, in the UK with respect to road inquiries. For an experimental period, library facilities are to be placed at the disposal of participants in local inquiries, containing documentation relevant to the inquiry itself and to the methodologies used by the Department of Transport in evaluating road investments. Copying of documents will be permitted, although this service will not be free. These are moves in the right direction and, as an absolute minimum, should be repeated for energy investments. Whether they will go far enough depends on the strength of belief in the basic principle enunciated at the beginning of this chapter and on an assessment of the likely costs and benefits of the wider dissemination of information.

In the energy field, more information has emerged in recent years about energy planning than hitherto. The Government has also promised a wider debate, as we saw. The same White Paper (Cmnd 6820, 1977) refers to the Government's encouragement of 'a wide public debate', to the need to reach 'a reasonable degree of public consensus on the acceptability of nuclear power' before embarking on any large-scale nuclear programme, and of promoting 'the fullest practicable public understanding of both the benefits and risks involved'. Certainly, a number of important documents now exist on which to base some understanding of energy policy in general.[6] It remains open to serious question, however, whether these documents reach more than the motivated public who would seek documentation anyway and whose activities have, in the event, actually led to the release of more information. Essentially, these are the professional 'cause' groups, on whatever side of the argument, and the academic researcher. This leaves the vast part of the public untouched by such documentation, for several reasons. Access to such documents requires an act of seeking out information in the first place, i.e. simply finding out what publications exist. The average householder is not on the mailing list for government publication lists. The average householder does not read the serious daily newspapers which might carry reports of the documents (even then occasionally omitting the full title and reference number to facilitate ordering). The average householder will not know how to order a government publication – few booksellers are HMSO agents; only a few towns have HMSO shops. In short, the very act of seeking information *about* information assumes a degree of articulation not possessed by the vast majority of persons. This does not imply that the public at large is incapable of coping with the information once received, but simply that knowledge of channels of communication is itself an acquired procedure.

To base an informational policy on documentation which will be released through these channels therefore tends to be self-defeating. It effectively assumes that what the information-generator knows about communication sources is known to everyone else, and that is simply not the case. Essentially, information has to be taken *to* the public. It should not be for the public to initiate the process of *acquiring* it. This simple distinction makes it all the more worrying to find statements to the effect that a great debate *is* taking place about energy policy. Such a belief is an illusion. What is taking place is a debate between professional elites. Interestingly, neither can claim, though both do (if we consider the nuclear debate alone), to represent public opinion. On the one hand there is now an open admission that the public have not been informed and hence it is illogical to suggest that the public 'support', say, the programme for nuclear power outlined in the Government's 1978 Green Paper *Energy Policy*. On the other hand no anti-nuclear group has tested public reaction to their own views. Statements to the effect that 'in order to help public understanding' documents on radioactivity research will be produced in simple language (Cmnd 6820, 1977, para. 41) are to be welcomed, but become all the more remarkable for their naïveté when one learns that such summaries are to appear in the Department of the Environment

'Pollution Paper' series, available through HMSO and therefore subject to all the criticisms above in terms of the public reached by such documents.

Of course there is a highly suggestive argument that the public at large do not care one way or the other about energy policy, or rather that they implicitly support government policy whatever that might be. The reasoning to support this view is varied. First, it is often suggested that if the public did not support such a policy they would vote against the Government at times of election. Second, it is suggested that their implicit votes are recorded by the fact that they do not seek information when it is available. Third, if they even wanted the information second-hand they would not read newspapers that failed to carry reports of the relevant debates. Fourth, it is argued that, silent though the majority are, they support whatever energy policy there is because they are capable of making the logical step of seeing that, without energy increases there can be no increase in GNP and their support for economic growth is itself proven by public behaviour.

Let us suppose that these arguments, or any one of them, is correct. Do they have implications for informing the public about energy and other issues on a wider basis? The fact is that they surely do not affect the argument that the basic axiom is not served unless wants and demands are known to decision-makers. We argued earlier that an informed demand is consistent with the basic responsiveness argument, while an uninformed demand is less consistent. If, in releasing more information, the result is to discover that the public at large is indifferent to energy policy then the debate between professional cause groups and the policy-makers can be properly construed as a 'public' debate. Elementary social choice theory tells us that we have recorded what the public wants if we record votes of indifference on behalf of the majority and votes one way or the other for the remainder. The important issue is to find out whether the majority vote is one of indifference. What social surveys we have suggest the majority vote is in favour of existing policy (White, 1977). But White also concluded that the public were *not* informed of 'the technical pros and cons of nuclear power'. Social surveys based on an informational context that is less than desirable do not really tell us very much except that on the basis of what information there is people vote in a particular way.

Nor is it the case that there is much force in the arguments for supposing that people support one policy rather than another. In the United Kingdom context energy policy does not divide political parties, so that it is difficult to see how the public can vote a party out of power on the basis of a disagreement with its energy policy. In recent years the exception to this has been the apparent espousal of the environmentalist cause by the Liberal Party. The argument about political division therefore has some limited validity in the context of minority government rule such as has been witnessed in the UK in the last few years. But it is fair to say that this stance on the part of the Liberals, itself an extremely small party, is of recent origin and it is unclear just how far it accords with national policy and therefore how far it is a permanent feature of UK political life.

The fact that the majority read newspapers or seek media outlets which do not record energy debates (or most major policy debates, be it over Concorde, reactor programmes, road programmes or whatever) is not in itself support for the view that they implicitly support whatever policy there is. Only recently has it been accepted that information needs to be 'comprehensible to ordinary people' (Leitch, 1977) in such domains. The cost-benefit model used to evaluate road schemes is hardly a news item of interest to the ordinary citizen, but a simplified exposition of it might be. While the media do often make attempts to simplify for the public, the process is unreliable and still tends to be confined to the more 'serious' media sources. There are dangers in simplification and it is clear that the process can go so far and no further. Quite how that balance is determined is difficult to say, but difficulty is scarcely a reason for not attempting it. It is fairly evident that the 'ordinary people' do not understand the elements of a nuclear fuel cycle because little attempt has been made to simplify it for them. In the nuclear debate, considerable effort is now being made by the Atomic Energy Authority and others to release material suitable for schools, local associations, public meetings and so on. Inspection of such information indicates that it is hardly 'neutral', nor would one expect it to be. A simple concept of fairness can be employed, therefore, to suggest that if information is to be presented in a non-neutral fashion those who dispute the views of one side should similarly release information. Since the information from the industry side is free of charge (and hence produced at the taxpayer's expense) so should that of the opposing groups be. In turn this suggests the taxpayer should subsidise an opposition which is not elected. There is clearly a dilemma involving a principle of fairness on one side and legitimate uses of public money on the other.

This dilemma may be best resolved by attempting the provision of information centrally which, where scientific dispute exists, actually records that dispute. Thus, the costs of generating electricity from power stations are disputed. The relative cost of, say, coal and nuclear stations would seem to be a relevant issue for public information. The hazards of low-level radiation are disputed and hence again the alternative views would seem to be a matter of public information. What is not legitimate is the presentation of *one* view. There is nothing odd about an informational programme containing more than one view and a debate on the views expressed. It matters again that such information be presented in terms comprehensible to the layman.

The view that implicit support for current energy policy exists because of a realisation that that policy is consistent with economic growth has some force. But it is important to note that clear statements of the costs and benefits of an energy policy do not exist. Nor, as yet, do clear statements of whether a given economic growth rate can be achieved by some other energy policy. In the former case it is important to 'sketch' scenarios of a future in which specific energy policies occur. These should include all the possible implications of those scenarios, economic, social and political. Such an exercise is taking place within the Department of Energy and it will be valuable. It would be useful to see others undertaking the same exercise. No energy policy will be systematically without costs. A nuclear programme has implications for civil liberties if nothing else. A programme with, say, limited nuclear power and a heavy insulation and conservation programme, may itself require increased regulations (e.g. on building design) or changes in energy prices through specific taxes. A programme of non-nuclear power may entail fairly complex implications for personal mobility if source and end-use are to be more evenly matched. The list of considerations is endless. That the dominant ones should be stated in documentary form is a clear prerequisite of an informational requirement for participation. Perhaps one should add that growth itself has costs and benefits and these might be rehearsed in a similar document.

While existing UK energy policy is based on the idea that alternative energy sources and conservation have a limited role to play in policy, other scenarios do exist in which it is suggested that given economic growth objectives can be achieved by policies which entail only limited nuclear power and certainly no fast reactors (see Chapter 7). It is also right therefore that these alternatives should be presented.

Will an information programme on energy policy be counterproductive? It will perhaps add to delays in given programmes if the information generates wants and demands which in turn become issues which require debate. But that is surely a price of democracy and it is unclear whether, if the information programme was begun immediately, it would hinder policy more than marginally. Certainly, the Green Paper on *Energy Policy* does not suggest there is urgency (see para. 10.17) at least as far as nuclear power is concerned. It is less obvious that there is no need for urgency as far as the *whole spectrum* of energy technologies is concerned. But since there are unlikely to be differences of opinion on many of them, this may not matter.

Nor is there any particular reason to suggest that an information programme would produce support for one side of the debate rather than the other. Nelkin (1977) reports the interesting finding that study of information programmes on nuclear power in Sweden 'indicated that more access to information may in fact increase confusion and conflict'. However, Nelkin's view is that part at least of this outcome was due to a failure to distinguish political issues from technical ones. This is almost inevitable in one-sided presentations and is a usual feature of existing literature. It is to be hoped that the presentation of more than one view would help to make the political issues explicit. WPI showed clearly how political issues can be brought out by adversarial process. It is largely for this reason that what is required in terms of institutional frameworks is separate processes of public consultation, one at the level of the national agenda and one at the local level. Technical and political factors will undoubtedly merge in both cases in practice, but securing some involvement on the political issues will serve to separate out the two factors as far as is possible. The danger, as Nelkin notes, is in confining both aspects to *one* process, in the UK's case to public inquiries.

Nor should anyone expect an information programme and a consultative programme to result in 'resolution' of conflict. This point has already been made. What is being served by the modified institutional system is the basic axiom of democracy. In the last event decisions must be made one way or the other.[7] That they should be made with the wider public participation promised by the UK Government is the requirement. An informational programme is part of letting people know what their opinions are and finding out what people want, not a mechanism for securing consensus, although the latter may, hopefully, result rather than the outcome reported by Nelkin for the Swedish experiments.

A final component of an informational programme is access to the background documents used by Government in evaluating and determining policy. That some progress has been made has already been noted. It remains the case that the United Kingdom Government has not accepted the view that there should be a statutory right of access to information as under the Freedom of Information Act in the USA. In the energy field in particular the debate that is occurring takes place largely in a context of suspicion. Those who seek to criticise energy policy, especially its nuclear component, see Government and the civil service as conspiring to withhold evidence and to 'cover up' events such as accidents in nuclear plant in order to make their own case as sound as possible in terms of its public image. Those within the 'establishment' often see the opposition as having sinister motives, hiding true motives behind a 'cause' such as antipathy to nuclear power. Two apparently simple ways to reduce this suspicion and thereby reduce the potential for conflict exist. One is to bring the two groups together. In the energy field this has been happening increasingly and represents a refreshing departure from earlier practice. Sensitivities remain but they would, on the evidence of WPI for example, be reduced, not increased, by more personal contact rather than less. Second, open access to background documents should permit the public to find out how thinking on a policy is determined. It might even improve that decision-making process. At the moment the public's chances to influence policy *in the making* arises only in respect of personal response to Consultative Documents (the 'Green Papers') or through being involved as an external consultant, a chance afforded to a limited number of persons. Moreover, a consultant may not be willing to criticise severely, even if he feels the internal documents deserve it, if his own career is partly dependent upon further consultancies. All in all, therefore, there is limited scope for influence. Yet, as Nelkin (1977) notes, the whole issue of public participation only works if participation occurs at the early stages.

The suggestion here is that a Freedom of Information Act should formally exist and should extend to public enterprises. The Labour Government has stated that it has an 'open mind' on the issue (Cmnd 7285, 1978). Its doubts about such an Act have been expressed as (i) that such an Act would be costly; and (ii) that it is redundant in face of the existing scrutiny of the executive by Parliament. The cost argument has limited validity, however, if one considers the chances that open access might prevent costly decisions from being made mistakenly. As an example, it is very arguable that more external opinion on Concorde might have led to it being cancelled, the right decision in the view of many at the time and clearly the right view with the benefit of hindsight. The second argument is weak in that the main reason for the growth of extra-Parliamentary institutions has been the fact that Parliament cannot cope adequately with all the tasks that would otherwise have to be presented to it. This is not a matter of saying it should not decide: it remains the final arbiter. It is a matter of saying that outside bodies can serve immensely valuable investigatory and advisory functions. In any event, the fact that advocacy of a Freedom of Information Act has wide support in Parliament, across parties, is indicative that even the scrutineers doubt their capacity to play the 'watchdog' role to the full without such access.

The case *against* a Freedom of Information Act must however be acknowledged. There are problems of defining which documents would come within the remit of such an Act. The USA Freedom of Information Act exempts documents relating to national defence and foreign policy, internal personnel rules and codes of practice, trade secrets, internal documents which would not be available on discovery, information relating to personal privacy, files relating to investigations for law enforcement purposes, some banking documents and data on geological information which could have national or trade value (e.g. oil wells), and other matters exempted by statute. Whilst accepting the problems of defining which documents should and should not be classified, there is much in the suggestion that a similar classification be adopted in the UK, with government departments (and, in our view, public enterprises as well) drawing up lists of other documents, by type, that they would wish to see classified and which could be so classified if Parliament approved the list (Outer Policy Circle Unit, 1977).

The main argument against such an Act is that it makes civil servants who are already reluctant to talk and say what *they* think even more reluctant to do so for fear that it will appear in a document which would automatically be unclassified. Many internal documents are initial thoughts, interim drafts, the posing of ideas rather than considered, final pieces of work. It is easy to see that their release through right of demand could generate more misunderstanding than anything else. Equally, items as small as internal memoranda on draft, interim and final reports can be and are accessible under the US legislation and it is far from clear that *less* has been committed to paper than would otherwise have been the case. Moreover, decisions have to be made on *some* basis and there is value in having access to the documents which Ministers would regard as being those that they have relied upon most, even if other documentation were not to be available. The difficulty with this kind of classification, however, is that it permits public access at a stage in the decision process when it is too late for any influence to be exerted. None the less, the distinction exists in the Swedish Freedom of the Press Act, where drafts and memoranda are excluded unless registered and filed at a public authority. The experience is that memoranda, etc., are in fact committed to paper and registered and that verbal communication does not take the place of documentation, which is the main fear of those who oppose Information Acts.

In so far as some documentation must exist for a decision to be made (if it does not exist, this would also be revealed by freedom of access and worthy of comment in itself!) the proper comparison is between the system as it is now in the UK, whereby extensive documentation is not released to the public, and a system under an Information Act regardless of whether such an Act 'silences' or 'puts underground' background material in respect of decision-making. If we take the worst scenario and imagine that civil servants will simply not document their thoughts, or will document reports which do not reflect their thinking, the question has to be asked whether this is worse or better than the existing system? The view taken here is that it would be better, simply because access to documentation that explains a decision is better than no access at all. As noted above, it is arguable that the undesirable effects of such an Act would not in fact come about at all. There is, for example, no evidence of this in the USA or Swedish examples.

FUNDING THE OPPOSITION

The main concern has been with procedures which attempt to embrace and inform a wider public than that so far involved in the so-called 'energy debate'. The suggestions on information have also largely been dictated by this consideration, although it seems fairly clear that a Freedom of Information Act would be most used, at first at least, by those who participate now in the debate and not by the general public. However, such an Act would also permit much wider coverage of issues by the media and this should readily extend general *public* information. One or two issues relating to participation by the 'cause' groups and interested individuals now require brief discussion. The first of these is the issue of funding objecting parties.

As it stands the law in the United Kingdom recognises only one group of statutory objectors (see Chapter 5). In compulsory purchase order cases such persons are entitled to have their costs of attending a public inquiry met if their objection is upheld (provided they have behaved 'reasonably'). Others are not.

If some financial assistance to these groups could be made available, it would serve two functions. First, it would assist in having a full debate and thus serve the public interest. Second, it would remove a potential argument that objecting parties could put, namely that the disparity in financial capability is such that objecting parties cannot take part in more than one such inquiry and can legitimately therefore (in the non-legal sense of 'legitimate') resort to other tactics.

A further fact requires some reflection. In many instances, the appellant at an inquiry will be a public body. The costs they incur are costs borne directly by the public. We take it that there is an auditing of such expenditures to ensure that they have been spent in proper fashion. The issue is, however, whether public money is being spent on only half the case. There would be no inquiry if it were not the case that the relevant Secretary of State considers there is more than one side to the argument (leaving aside the cynical view that inquiries are there to appease objecting parties while reaching the intended conclusion anyway). It seems logically odd therefore that public money should be spent on the institution of the inquiry, and, if the one party is a public agency, on funding one side of the argument but not the other.

It is perhaps worth adding that the problem of mistrust, if not distrust, which is very widespread among even the most intelligent and informed of the objectors at WPI is exacerbated by the refusal to assist them financially. An effect of even limited funding would be to confer some legitimacy on objectors, to raise the status of intelligent dissent, thereby encouraging participation in orderly, democratic procedure. It would also weaken the position of those who are so distrustful as to claim that dissent is tolerated but not taken seriously and that public inquiries of any sort are no more than 'a sham', a view which is also widely voiced. This is not to hold a pistol to society's head, but rather to suggest that informed criticism publicly expressed is a valuable ingredient of a democratic society.

Given the complexity of our society and the issues involved it is hardly surprising that criticism costs money, but to deny this resource to the would-be participant is contrary to an informed and efficient democracy. The *principle* of funding objecting parties, statutory or otherwise, should therefore be accepted. Official resistance to the principle is not total – see, for example, the UK Government's response to the suggestions made in Dobry (1975). Also the Secretary of State for the Environment stated in Parliament on 6 March 1978 (in the context of the Windscale Inquiry) that 'I shall consider further – perhaps for future inquiries – the implications of a protracted inquiry of this kind and the cost that it imposes on those taking part' (Hansard, 6 March 1978, col. 991). This contrasts, however, with the refusal to consider funding for objectors at road inquiries (Cmnd 7133, 1978).

Exactly how such funds should be administered, if the principle is accepted, is a source of debate, although experience elsewhere suggests that the problems are not insurmountable.

Various possibilities have been discussed in Dobry (1975) and in Lock (1977). We would suggest that parties wishing to object and who wish to seek financial assistance should be allowed to submit a claim both prior to the inquiry, and after the inquiry in the event that unforeseen costs occur (and especially if they misjudge the length of the inquiry). Such submissions would in any event also be better followed up by a statement, certified by an accountant, as to how the money was spent. In deciding on funding concern should be with the standing of the objector or objecting group and with the likely content of the submission. Where submissions clearly overlap, objectors might usefully be asked to contact other objectors to see if their submissions might not be integrated to save both time and money. This would seem to be a valuable function for an inquiry secretariat. It was the procedure followed by the Fox Commission in Australia.

By allowing applications for funds, those groups who wish to remain independent of central funding can do so. There is of course no hard and fast line to draw between who is deserving of funding and who is not. Inquiries are notorious as contexts within which individuals seeking some form of grandeur can, sometimes, achieve it regardless of whether they care about the issues in question or not. This seems an unavoidable risk, however.

One other problem concerns the fact that some objecting groups may have other sources of funds. Unless there is a means test, it may well be that an initially endowed group or individual could secure undue advantage from such a scheme. We doubt if a means test is feasible or desirable. We would suggest that applicants for funds should be asked to state whether they have other funds available. Given the state of knowledge about how campaigns and inquiry appearances are funded – i.e. the fact that funding tends to be generally known of by anyone interested enough to find out – there would be some incentive for applicants to be honest in their response. Further, we would suggest that levels of funding and spending by *all* parties to the inquiry be published during any inquiry so that representations might be made by other parties if they so wish. This opens the possibility of 'in-fighting' between groups and individuals but this kind of public knowledge may be necessary to prevent abuse of the provision.

Lastly, the issue arises of what funding should be for. It seems clear that, if the proceedings take place before a national Commission, lawyers' fees will not be relevant since lawyers will not be involved. If the inquiry is through a local inquiry or planning inquiry commission, then lawyers can be involved and their fees would be a legitimate expense. Travel to and from the scene of the inquiry would be a legitimate item of expense. We are less sure about any forgone wages or salaries, which could be a substantial item. Where an objector does secure an agreement with his or her employer to have leave of absence, unpaid, this can be readily verified by the Secretariat of the Commission or Inquiry and the case may then be judged on its merits.

If there is concern over the cost of funding it might be possible to establish a guideline for overall cost and conditions under which it might be exceeded. (We note in this connection that BNFL estimate their direct expenditure on WPI at $£\frac{3}{4}$ m.)

INDEPENDENCE OF ASSESSMENT

A final requirement for an efficient procedure is that those assessing the debate, whether at a local inquiry or a national inquiry, should be independent. Politically, it would in any event be an unlikely strategy for any Government to appoint persons who are not independent, since this could only be seized upon by objecting parties who can then 'legitimately' claim that the process was unfair. In this way, no inquiry could be regarded as politically acceptable. None the less, there are problems in any form of inquiry into energy matters. This is most likely in the nuclear context where feelings run high, for good and bad reasons. The problems arise in that technical assessment will require expertise in nuclear matters and in the United Kingdom it is often alleged that there is a difficulty in securing persons independent of the industry. Almost by definition they will have been employed by, or have been indirectly involved in, the industry. In itself this is no reason to suggest lack of independence, not least because assessors, inspectors at local inquiries or members of national commissions are eminent persons in their own right. But independence, like justice, has to be seen to exist, and this poses a genuine difficulty in cases where established policy in government and industry favours one position because it then becomes very difficult to find any one assessor (let alone assessors) who may be fairly deemed independent in the sense of being free of prior conditioning to the prevailing view. Even where such people can be found, unless they are *believed* to be independent by all parties to the inquiry, their appointment serves only to add to the problem of mistrust referred to earlier.

An alternative to seeking independent persons would be to seek partisan persons and to ensure that an inquiry or Commission is composed of opposing assessors. This 'dialectical' composition would then either result in an agreed assessment of the evidence or two reports. By virtue of our argument that we doubt if nuclear conflicts will be 'resolved' at inquiries, the former outcome seems unlikely in the nuclear context. The latter, while initially looking like an odd suggestion, is not of course particularly unusual in a system which is already used to majority and minority reports. Further, the existence of two reports could be held to assist public involvement and to ensure that the ultimate arbiter, whether it is a Secretary of State or Parliament is apprised of both sides of the picture. Typically this philosophy would run counter to what is expected of a local public inquiry where a single inspector (sometimes with assessors) sits in judgement precisely because the Minister seeks unequivocal advice (which he may nonetheless reject). None the less, the suggestion is worth more attention than has perhaps been paid to it.

ADVERSARIAL CONTEXTS

The context of an inquiry process can vary from the extreme of being barely distinguishable from a courtroom to the kind of fairly informal procedures used by Parliamentary Select Committees or national Commissions. Cross-questioning must self-evidently exist. But whether adversarial contexts should be encouraged is the subject of debate. It has not been accepted in the case of road inquiries, where the Government has called for *less* of a 'courtroom atmosphere' (Cmnd 7133, 1978). The argument against such legalistic proceedings is that they inhibit participation and we have suggested that considerable weight be placed on participation at all levels. Anyone who has been involved in inquiries will be aware that they are not pleasant experiences for other than those who make a 'profession' out of such appearances. Consequently, even technical experts can be deterred from attending. The 'public at large' has even less incentive to give evidence since the entire context is beyond their norms of articulation. Where adversarial contexts can be avoided, therefore, it seems sensible that they should be.

The argument in favour of adversarial proceedings is a powerful one in circumstances where there is any reason to believe that information will not otherwise be volunteered. Again, noting the suspicions surrounding the nuclear debate, this would seem to be relevant to that context. It may even be true of all energy debates. As such, adversarial processes can have a valuable function in generating information. This was the case at the Windscale Public Inquiry, which used an adversarial process (indeed, witnesses were placed under oath). Adversaries need not of course operate through legal representatives. A standard complaint at many public inquiries is that lawyers frequently 'intervene' between the two parties, losing points of importance through lack of technical knowledge. This complaint is most frequently heard from technical witnesses who would prefer to cross-examine persons in a similar capacity in their own language. The problem here is that, say, the examination of one radiation specialist by another, or one economist by another, risks the use of jargon which 'loses' the remainder of the audience. Another view is that legal training is devised to elicit information and laymen may not be so efficient at this process.

Whether lawyers should be present or not, there is a case in energy matters for an adversarial process. We may note, however, that a Freedom of Information Act extended to public corporations would in itself assist in the removal of the need to approximate inquiries to court cases since there would be an obligation on the part of the energy agencies to release information on request. This is one spillover benefit of such an Act. In the absence of such an Act, however, adversarial processes seem to be needed.

SUMMARY OF CRITERIA FOR EFFICIENCY

We may now summarise this lengthy discussion of what constitutes the requirements for a procedure to be 'efficient'.

Efficiency must be measured against an objective. The fundamental objective suggested here is that of the responsiveness of Government policy to expressed public wants. Any inquiry process must therefore make more effort to discern what public opinion is or where the national interest lies. That function is best served by a twofold procedure – one at the national level and one at the local level. The national procedure has the function of investigating, evaluating and informing. The local procedure comes into play only if local residents consider there is an issue at stake. If there is, a local inquiry can legitimately serve the function of discerning the local wants and evaluating the impact of a single investment on the locality. Where an investment has both local and national dimensions it is far from clear that the local inquiry *alone* can be considered efficient as a means of reaching an advisory decision. As we shall see, the device of a planning inquiry commission exists (though never put into practice) for this very purpose. The requirements for efficiency are therefore

(1) that whether local or national, the advisory or investigatory procedure be apprised of the maximum amount of information relevant to the decision.

(2) that local procedures be apprised of local opinion;

(3) that national procedures culminating in Ministerial or Parliamentary decisions be apprised of general public opinion in so far as this is possible;

(4) that requirements (2) and (3) are best served by an informational programme designed to present the pros and cons of an investment or policy in comprehensible language.

(5) that a Freedom of Information Act would assist in securing the fundamental axiom of democracy stated at the outset.

(6) that objecting parties at local or national inquiries be funded, subject to qualifications.

(7) that those assessing evidence should be independent or that the assessing body be composed of members with counteracting views.

(8) that, in the energy field, adversarial proceedings be used unless a Freedom of Information Act is in existence.

In discerning wants it should be clear that any procedure will be discerning *values*. We return to this issue in the case study of Windscale since, in the energy context, and indeed in the general context of development, it will be seen to be important.

Subsumed within these requirements is the other major consideration: the search for a 'correct' statement of facts. In many cases this may be a simple process of establishing what is or what is not the case compared to what people believe to be true (this was one of the most effective features of the Windscale Inquiry in some respects). In other cases it will involve a balancing of technical evidence since, as we noted, the technologies that are now under consideration, and which will increasingly be under consideration, are not necessarily proven either in themselves or in terms of their environmental consequences. Ultimately, however, an advisory procedure has to report on the *desirability* of a policy or investment, national or local, and hence it must report on what people want or explain why those wants should be overruled. The biggest danger in the existing system of institutions in the UK is that advisory procedures do not have mechanisms for testing public want. It is because of this that procedures require modification. The need for efficient decision-making procedures will increase. In the first place, recent years have witnessed the emergence of a new ethic which is opposed to the values which underlie increasing national wealth through the exploitation of limited natural resources and a finite environmental capacity to assimilate the waste that necessarily results from the creation of wealth. How widespread support is for this new ethic is difficult to say. Casual inspection suggests that the vast majority of persons want more of what they now have and especially want things that they do not have. But it would be wrong to state this categorically and even more wrong to assume that the proponents of the new ethic are without voice or even power. Second, even if incomes rise in line with traditional goals for economies, there is evidence to suggest that the demand for environmental goods increases with those rising incomes (this is not the same thing as saying such goods are 'elitist' – see Pearce (1978)). If this is correct, conflict must arise, especially if this increasing demand is associated with a decreasing availability of environmental goods through the very process of permitting industrial investment. We can therefore expect conflict both to exist and increase in frequency and perhaps in duration. Third, we have the growth of extra-Parliamentary institutions through which those embracing a different ethic from one based on expansionism can express their views. Fourth, there may be some truth in the view, often stated, that the maintenance of a growth-oriented economy requires larger and larger investments to sustain that growth (since in absolute terms the increment to national income or wealth gets larger and larger) and that this requirement has placed strain on the capability of technology to keep pace. If that is true, we must expect some future investments to be into technologies which are uncertain in their structure: that is, there may be a temptation to commit the nation to a technology which is comparatively untried. Of course, all new technology has unknown elements. The issue is whether the cost of making a mistake plus the cost of the technology exceeds the benefits from it, where the cost of the mistake is probabilistic. This way of thinking of things is not fanciful or even unduly cautious: recent experience with drugs that have had unforeseen side-effects greater than the benefits of administering the drugs should be sufficient to illustrate the argument. Fifth, the demand by the public, or some section of the public, for participation is so well established as to require accommodation.

If the preceding analysis is even remotely correct, it matters a great deal how we deal with new investments which have national attributes. As Gladwin puts it, 'It is not environmental conflict itself that is dangerous, *but rather its mismanagement*' (1978, our emphasis). In short, the need for efficient decision-making units will grow.

NOTES

1. In terms of Arrow's theorem (1963) this condition is referred to as 'non-negative association' – a social ordering of events must respond to a discernible change in the ordering by individuals. At its simplest level it amounts to saying that individual preferences should count.
2. This requirement does not extend to all internal papers relevant to a policy decision. As such it is not consistent with the obligation placed on Government in the USA Freedom of Information context. The relevant UK Government document states that 'the working assumption is now that, once Ministers have reached their conclusion on a particular major policy study, associated factual and analytical material will be published.' See Home Office, *Reform of Section 2 of the Official Secrets Act, 1911*, Cmnd 7285, July 1978 (HMSO, London) para. 42. The text of this chapter explains why this statement is held to be less than fully consistent with the informational requirements for efficient decision-making.
3. Technically, the Krutilla-Fisher model requires that the benefits of environmental preservation be compounded forward at some rate of interest which reflects the relative price change of unique assets *vis-à-vis* development benefits. This relative price change will be positive (*a*) because natural assets are less and less available through time, and (*b*) because development benefits are subject to technological change which renders them less valuable in future as substitutes emerge. Interestingly, the model preserves the standard neoclassical assumption of a positive discount rate and we have argued that this may not be legitimate.
4. Interestingly, *local* inquiries have taken place in which alternative sites have been proposed during the inquiry. This is, however, widely regarded as 'illicit' procedure. At the inquiry into the Bacton North Sea Gas terminal in 1967 amenity groups opposed the Bacton site but could not agree on which alternative sites to propose to Shell, the intended developers. Meanwhile the Inspector himself was collecting information about two alternatives *he* had put forward. Norfolk County Council requested the Minister to widen the scope of the Inquiry to consider yet another site they put forward and the Minister agreed to this. A fourth site was proposed by the National Parks Commission. Shell and the Gas Council looked into the feasibility of all of them.

   This raises another interesting point, which is whose duty it is to bring alternative sites forward at a public inquiry. In *Rhodes* v. *Minister of Housing and Local Government and Another*, December 1962, the judge found that the question of alternative sites first of all *depended on the nature of the application.* Secondly it was found that it is not incumbent on the Minister to do more than consider the facts put before him by an inquiry, including whether it has been shown by objectors that an alternative site is available. But as we have seen, in the Bacton inquiry a request was made to the Minister to permit the introduction of another site for discussion; alternative sites being put forward by the objectors and even the Inspector.

Prior cases would seem to suggest that it would be possible for objectors to introduce *suitable* alternative sites for discussion where power stations are concerned, because of the nature of the development. None the less, the prevailing opinion does not support this view.

5. To add to the confusion, in 1978 the UK refused to back a call from the European Community for a fast reactor programme. At one and the same time, the European Commission, empowered in March 1977 to issue loans on behalf of the European Atomic Energy Community (Euratom), issued a loan of 66 million EUA (European Units of Account), or about $80m, to the Super Phénix fast reactor in France. In effect, then, the UK as a Community member is assisting in the finance of the French fast reactor whilst simultaneously promising an inquiry into a UK fast reactor, an inquiry that only makes democratic sense if at least the *possibility* of not building such a reactor is an outcome of that inquiry. To put it another way, the Euratom loan could be seen as a commitment to fast reactors. To add to the suspicion, this information was not released by Euratom but was discovered independently by an environmental group, the European Environmental Bureau, based in London.

6. These are, notably, Cmnd 6618 (1976); Montefiore (1977); Royal Institution (1977); Select Committee (1977); Parker (1978); Cmnd 7101 (1978); Department of Energy (Energy Papers Series); Energy Commission (Energy Commission Paper Series).

7. As instances of the complexity of advancing views for change in a context of suspicion about the motives for change we may cite just two reactions to the Interim Report produced by the authors in July 1978. The Energy Daily, 18 July 1978, reporting on an abbreviated form of the Report, stated that the authors advocated that 'the institutions making decisions must be modified to reflect those opposing the decisions', and commented that this was 'the kind of thinking that has produced the nuclear moratorium in the United States . . .'. The *Financial Times*, 14 July 1978 reported that the paper' . . . proposed for the forthcoming fast reactor inquiry a formula which, at first sight, will ensure that a clear-cut decision always remains beyond reach'. The substance of what was *actually* said, as in this chapter, was, however, (i) that institutions should be modified to encompass a wider public participation than the opposition 'elite'; and (ii) that decision procedures must have built-in guarantees that a decision, one way or the other, must be made within the time-horizon of permissible delay, which, in the case of CDFR was put at 'up to 2 years'. The *Energy Daily* correspondent also seemed unaware of the fact that delays in the USA have arisen through the use of Court procedures, whereas no such procedures are open to non-statutory objectors in the UK.

REFERENCES

Arrow, K. (1963)   *Social Choice and Individual Values*, (2nd edition, Wiley, New York).

Dorfman, R. (1973)   'Conceptual Model of a Regional Water Quality Authority', in Dorfman, R., *et al.*, *Models for Managing Regional Water Quality* (Harvard University Press, Cambridge, Mass.).

Haefele, E. (1973)   *Representative Government and Environmental Management* (Johns Hopkins University Press, Baltimore).

Henderson, P. D. (1977)   'Two British Errors: Their Probable Size and Some Possible Lessons', *Oxford Economic Papers*, July.

Burn, D. (1978)   *Nuclear Power and the Energy Crisis* (Macmillan, London).

Breach, I. (1978)   *Windscale Fallout* (Penguin Books, Harmondsworth).

Rowen, H. (1976)   'Policy Analysis as Heuristic Aid: the Design of Means, Ends and Institutions', in Tribe, L., *et al.* (eds), *When Values Conflict* (Ballinger, Cambridge, Mass.).

Downs, A. (1958)   *An Economic Theory of Democracy* (Harper & Row, New York).

Sen, A. (1967)   'Isolation, Assurance, and the Social Rate of Discount', *Quarterly Journal of Economics*, vol. 81, March.

Harsanyi, J. C. (1953)   'Cardinal Utility in Welfare Economics and in the Theory of Risk Taking', *Journal of Political Economy*, October.

Marglin, S. (1963)   'The Opportunity Costs of Public Investment' and 'The Social Role of Discount and the Optimal Rate of Investment', *Quarterly Journal of Economics*, March, June.

Krutilla, J., and Fisher, A. (1975)   *The Economics of Natural Environments* (Johns Hopkins University Press, Baltimore).

Select Committee on Science and Technology (1977)
Third Report, on *The Development of Alternative Sources of Energy for the United Kingdom*, Vols 1–3 (HMSO, London).

Montefiore, H., and Gosling, D. (1977)
Nuclear Crisis: A Question of Breeding (Prism Press, Dorchester).

White, D. (1977)
'Nuclear Power: A Special New Society Survey', *New Society*, 31 March.

Leitch, Sir G. (1977)
Report of the Advisory Committee on Trunk Road Assessment (HMSO, London, October).

Nelkin, D. (1977)
Technological Decisions and Democracy (Sage Publications, Beverly Hills).

Cmnd 7285 (1978)
Reform of Section 2 of the Official Secrets Act 1911 (HMSO, London).

Dobry, G. (1975)
Review of the Development Control System (HMSO, London).

Lock, D. (1977)
Planning Inquiries: Paying for Participation (Town and Country Planning Association, London, August).

Select Committee on Science and Technology (1978)
Third Special Report, Session 1977–8, 20 July (HMSO, London).

Cmnd 7133 (1978)
Report on the Review of Highway Inquiry Procedures, (HMSO, London, April).

Department of Energy (1978)
Energy Forecasting Methodology, Energy Paper No. 29 (HMSO, London).

Cmnd 6618 (1976)
RCEP – *Sixth Report, Nuclear Power and the Environment* (HMSO, London).

Cmnd 6820 (1977)
Nuclear Power and the Environment: The Government Response to the Sixth Report of the Royal Commission on Environmental Pollution (HMSO, London).

Royal Institution (1977)
Nuclear Power and the Energy Future (October, Royal Institution, London).

(Mr Justice) Parker (1978)
The Windscale Inquiry: Report (HMSO, London).

Gladwin, T. (1978)
'The Management of Environmental Conflict – a Survey of Research Approaches and Priorities', New York University (*mimeo*).

Cmnd 7101 (1978)	*Energy Policy: A Consultative Document* (HMSO, February).
Schroeder, H. (1974)	*The Poisons Around Us*, (Indiana University Press, Bloomington).
Nobbs, C., Pearce, D. W. (1976)	'The Economics of Stock pollutants – the Example of Cadmium', *International Journal of Environmental Studies*, January.
Easton, D. (1965)	*A Systems Analysis of Political Life* (Wiley, New York).
Pearce, D. W. (1978)	'The Social Incidence of Environmental Costs and Benefits', forthcoming in *Progress in Environmental Planning and Resource Management*, Vol. 2, 1979.
Darmstadter, D. (1977)	*How Industrial Societies Use Energy*, (Johns Hopkins Press, Baltimore).
Otway, H., *et al* (1978)	'Nuclear Power: the Question of Public Acceptance', *Futures*, April.
The Outer Circle Policy Unit	*An Official Information Act*, (London, 1977).

4 Existing Advisory Institutions I

GENERAL

Any analysis of the 'efficiency' of energy decision-making in the UK must begin with a description of what institutions exist and how they relate to each other. It must also consider whether they serve, or could serve, the following functions:

(i) the assessment of single investments which have 'local' attributes;
(ii) the assessment of single investments which have 'national' attributes;
(iii) the assessment of programmes.

Under (i) we also need to consider whether a local investment is necessarily 'wedded' to a specific location (as with, say, mineral exploitation) or is 'spatially mobile' and could be placed in one of several locations.

In reality, the UK would seem to have more committees and provisions for committees than she knows what to do with. There are *statutory committees*, i.e. those that are defined and provided for by Act of Parliament generally, such as the Planning Inquiry Commission, or specifically, such as the Coal Industry Commission set up under the Coal Industry Commission Act of 1919. Other committees or commissions may also be called 'statutory' because they are implied by an Act of Parliament which gives a Minister the power to set up a committee or commission. *Non-statutory committees/commissions* are set up under the Minister's conventional power or prerogative. Hence the system is flexible. In theory then, a commission or committee may be established to fit any particular need. With regard to the power of these various committees, commissions, working groups or whatever, there are in some cases only very fine distinctions which have grown out of precedent and tradition. Of course the remit given to a committee or commission coupled with the powers voted to that body by the Minister and/or Parliament can create greater distinction.

Commissions, etc., are administrative tools and hence have flexible procedures: they are not tightly bound by legal structures. Thus a Royal Commission's procedures and powers have been derived in an *ad hoc* fashion over a long period of time. We look at this first.

(i) Ad Hoc Bodies: The Ad Hoc Royal Commission[1]
A Royal Commission consists of:

(1) a Chairman.

(2) Members – flexible numbers.

(3) Secretariat: the size varies according to the needs of the Commission. There may be one or two secretaries depending on whether one or more government departments have been responsible for initiating the setting up of the Commission. Secretaries are usually members of the Civil Service, though not always.

(4) Assessors: numbers again vary according to the purpose of the Commission. Whereas virtually every Commission has a secretariat only a few have assessors. Assessors may be experts from within the Civil Service or without, but they are appointed, like the secretariat, by the interested government departments.

(5) Research Staff: some Commissions employ research staff, others do not, and this is often the Chairman's decision. The late twentieth century has seen the growth in the use of full-time independent research staff, recruited *ad hoc* from inside and outside the Civil Service. The Research Director may be appointed as an Assistant Commissioner so that he may attend committee meetings.

Depending on the problem in hand the Royal Commission may split into sub-committees headed by a member of the committee who has been asked to investigate a particular aspect of the remit. These sub-committees may co-opt members from outside, e.g. university academics or other persons with suitable expertise. All the members of the Commission may go on to sub-committees so that the parent Commission acts merely as a co-ordinating body, or it may be that there are only one or two sub-committees of the Commission. The secretariat co-ordinates the timetabling of the sub-committees and parent Commission.

The Chairman, with or without advice from the secretariat, determines the procedure of the Commission. Usually a preliminary meeting is held where literature and documents on the background to the problem are distributed by the secretariat to the members. This background information may well be given by the interested departments. At this initial meeting the remit of the Commission is discussed as fully as possible (another meeting may well be necessary). Usually the terms of reference are fairly uncomplicated and wide-ranging, as the Minister hopes the Commission will find its own direction and make its remit more concise. Sometimes a Commission may ask for the terms of reference to be made clearer or the Minister himself will attend initial meetings to discuss them. Usually the Commission reaches an understanding on the terms of reference indirectly via a debate in the House of Commons or through some other method of communication on what type of interpretation the Minister would prefer. In the course of the Inquiry, the Commission may ask for a formal amendment in its terms of reference (or they may be initiated by the Minister).

At the preliminary meeting the members discuss the kind of information they need and how they should go about obtaining it, e.g. by open invitation to the public in the press, restricted invitation, request for oral or/and written evidence, etc.; whether they should make visits to obtain information at first hand or undertake special investigations; frequency, dates, length of future meetings are discussed, as is the procedure to be followed at these meetings.

Meetings are usually held in London but sub-committees or the parent Commission are at liberty to decide whether or not to 'rove'. Oral evidence is usually taken as a 'back-up' to written evidence and although the Commission may choose to be represented by counsel, witnesses are usually not so represented. Nevertheless the Commission may use its discretion whether or not to allow witnesses to be legally represented. In general legal representation would be the exception to the rule.

Once all information and evidence that the Commission deems necessary is gathered, the report is drafted. Quite often this initial draft will be undertaken by the secretariat, or, if the Chairman feels this would imply some departmental influence, he will draft the report on his own, with some members appointed as a sub-committee or in a meeting of the whole committee. The report is discussed in committee and amended. Those who dissent may sign the report but also append some reservations, while other dissenters may refuse to sign the report and produce a minority document. The report is eventually presented to the involved Minister(s), who decide whether or not to publish it. It is then presented to the Queen. It is presented in Parliament as a Command Paper. There may be a formal statement made by Minister(s) in Parliament on the more important and contentious Commission reports. *The Government is not bound to implement any of the report's recommendations or to reach a decision on the report.*

There are some major inherent drawbacks to the Royal Commission as an advisory body. It is *ad hoc* (temporary), and its members are only part-time. Furthermore the Minister may totally ignore its recommendations, although a reaction of some sort is usually politically expedient.

As an *ad hoc* body, once the report is presented the Commission is automatically disbanded. Hence it has no power to give further advice to government on the implementation of recommendations and is also unable to review the situation at a later date once the results of its recommendations are seen. It is possible to recall a Royal Commission if the need arises, but continuity would be lacking.

The employment of unpaid persons as part-time commissioners means that only a certain type of member may be asked to serve, i.e. those who are able to take time off from their normal employment or those who have retired. Commissioners, therefore, tend to be older men and women who have made their reputations and perhaps have a decided outlook on life. Hence Royal Commissions tend to be conservative bodies.[2] There is something to be said for the appointment of persons who still have their reputations to make and payment of members could perhaps lead to Commissions being made up of those from a wider social background. How realistic this is is questionable, given that most members are appointed to represent given interests and have established reputations. Those not representing interests are generally expected to have some eminence in relevant fields of study.

Royal Commissions tend to be treated liberally with regard to costs, although on some instances they may be asked to justify expenses. For example the Commission may broadcast on TV or on the radio when issuing invitations to the public to come forward and give evidence; but we have no evidence as to what would happen if they requested a public information campaign of some sort prior to the issuing of invitations. At present costs are voted in a distinct vote in the Civil Appropriation Accounts. It is difficult, as Cartwright states,[3] to determine the degree of independence of the Royal Commission with respect to expenditure. But there would seem to be indications that an experienced Chairman who has knowledge of the workings of officialdom is able to obtain what he deems necessary. If the Commission saw the need for the production of independent cost-benefit analyses and environmental impact statements it is likely that these would be financed under the research criteria; but there would evidently be some departmental discussion about this and official opinion as to the viability of a course of action would no doubt filter back to the Commission via the secretariat. With regard to 'reaching the general public' the status and respect a Royal Commission commands means that they may promote public discussion by periodic press releases, especially if the press receive a piece of 'leaked', 'confidential' information which keeps public interest alive, focuses discussion and allays suspicion.

Although a Royal Commission is in theory commanded by the Queen, in practice the appointment of such a Commission is very much a Ministerial, departmental decision. Assessors, secretariat, Chairman, members are chosen by Ministers and departments, hence the Government exerts a strong influence on the composition of the Commission, so much so that the Minister may often be fairly confident in predicting the line or sort of views the Commission might take on the issues. But if Parliament suspected the Government of 'packing' the Commission it could certainly demand changes in membership, etc., when questioning the relevant Minister in the House of Commons. Should the choice of Commission members be made more 'democratic' by having the appointment of a Royal Commission subject to Parliament or at least the terms of reference of a Commission being discussed in Parliament and subjected to amendment as a matter of course rather than the exception? Cartwright feels that if such limitations were imposed on Ministers, they would resort to using less formal and less open kinds of committee, appointed without recourse to Parliament, and that it would be impossible for Parliament to forbid this type of action. It seems unlikely that all Royal Commissions warrant the discussion of their mandate in Parliament; but some clearly do, where contentious national issues are to be investigated.

A major drawback of Royal Commissions is that they have very little discretion with regard to the evidence they must receive. For example a Royal Commission on a nationally contentious issue is appointed and publicised. It must then invite evidence. If it restricts the evidence it is willing to receive it will be accused of bias. If it offers an open invitation it may be flooded with repetitive mountains of evidence that could take months of sifting. It has been suggested that Commissions should try and assess the value of evidence likely to be presented to them before they issue invitations. Would the public accept this action? Would inclusion of all the pressure groups in the invitation prevent them from complaining on behalf of the general public? It may be that the answer is a larger secretariat employed temporarily to sift, summarise and assess written evidence, with a commissioner in charge of the operation making the decisions on who to call to give oral and other evidence.

On the more positive side, Royal Commissions are left to more or less carry on with their work as they think fit. The Treasury produces a set of notes of guidance (unpublished) on certain general principles respecting the conduct of business. But procedure may be developed to suit the problem. Once the Commission is under way there is very little departmental involvement unless the Commission requests it.

'Success' in the end depends on the original appointment of Chairman and members, not only personalities but also composition. The Minister may elect to have either a Commission of experts, or representatives of interest groups, or an 'impartial' Commission. Each of the three different types will view the problem in different ways. The expert and representative Commissions would probably limit the amount of evidence they felt necessary to incorporate. They would have prior background knowledge of various problems and issues. A representative Commission would at least present 'open' bias but the members would need to be very flexible and open to some compromise. An impartial Commission would perhaps contain some expert and representative members but on the whole it would be a 'lay' body. It would therefore need more time to digest background material, need more assessors on technical subjects and have less concept of the type of evidence needed. But it may well be more publicly acceptable. In practice, the composition of Commissions is determined by a complexity of motives and pressures. The process of consultation is unclear and close scrutiny of membership shows that departing members appear often to nominate their successors.

Time is an important factor. Royal Commissions usually take from two to five years to reach a conclusion. The part-time nature of the body lengthens the time that could be taken. It may act in a speedier fashion on the request of the Minister. This, of course, may affect research projects and witnesses. As it is appointed in theory by the Crown a change of government will not affect its deliberations, but it is not perhaps the right vehicle for the making of urgent decisions. Certainly, in the past there have been substantial gaps between Reports of Royal Commissions, the implementation of recommendations, and even the debate of recommendations. Royal Commissions do from time to time on their own initiative or at the request of the Minister issue interim reports which test the acceptability of sometimes radical ideas and keep issues 'alive'. It is open to the Minister whether to adopt or at least begin to consider any of these interim recommendations before waiting for the final report.

Interim reports and final reports will have their dissenters, which may weaken the effect of the majority recommendations. Without dissent there may well be a woolly compromise which would be just as unacceptable if not more so. Dissent is to be expected with a large body of individuals who may never have met or worked together before.

Overall, despite the motives which might lie behind the setting up of a Royal Commission, its purpose is primarily to offer advice which it undertakes to give by obtaining information and formulating policy upon which it will propose specific action.

It is important that a Royal Commission 'keep its feet on the ground' when making recommendations in the drafting of its report. The recommendations must be 'practical' from the Government's point of view. The Secretary or any Civil Service member on the Commission would be a help in this respect. Obviously the point of 'practicality' creates a dilemma – does the Commission wish to be controversial and an embarrassment to the Government, or even to have its recommendations ignored; or, at the other extreme, to be very conservative and even ambiguous in its recommendations to the detriment of its own public image?

It could be said that the Royal Commission is flexible enough to consider energy policy and make recommendations and also to cope with planning questions. But the *ad hoc* Royal Commission is essentially a 'one-off' process so that to request that it make recommendations on general energy policy and its subsequent planning implications would seem to be absurd. It is too much to expect the one procedure to cope with.

(ii) Ad Hoc Bodies: The Departmental Committee
The Departmental Committee differs little from the Royal Commission in terms of structure and flexibility of procedure. Fundamental differences lie in the fact of appointment: the Royal Commission (in theory) is appointed by the Crown and is therefore not affected by a change of government, while the Departmental Committee is appointed by the relevant Minister(s) and may be discontinued if there is a change of government. (This has never happened in practice.) The Royal Commission has the power of subpoena, which the Departmental Committee does not, but neither committee has much difficulty in sending for people and papers. If such power was needed by the

Departmental Committee it could be voted by Parliament. The Royal Commission carries out research, the Departmental Committee rarely does so. The Departmental Committee reports more quickly and is less expensive (but at a price). Neither the Royal Commission nor the Departmental Committee need print evidence and research. In practice, most Royal Commissions do and most Departmental Committees do not. Hence it is very difficult to see which evidence the Departmental Committee is most influenced by.

The final report of the Royal Commission is usually a Command Paper while that of a Departmental Commission may be a Command Paper or a non-Parliamentary paper which may or may not be implemented by the Minister.

Departmental Committees have less status and prestige, get less publicity than Royal Commissions, and are perhaps less trusted or known by the public. In theory and perhaps in practice they are just as adaptable, open, participatory and impartial as the Royal Commission. Their members comprise academics, MPs, civil servants, etc., but the permanent head of the Civil Service gives advice on membership. The Franks Committee on Section 2 of the Official Secrets Act of 1911 (short title) may be cited as an example of the type of work a Departmental Committee could undertake. It was appointed on 20 April 1971 and adjourned on August 1972.

The Departmental Committee has most of the drawbacks of the Royal Commission and even more of its own. Like the Royal Commission it is *ad hoc* and part-time, but it has less prestige and its impartiality would be more likely to be challenged by the public. It receives less publicity than the Royal Commission and has in the past been used to debate more issues of 'national' importance than the Royal Commission, partly because of this anonymity. It is less expensive and time-consuming on average than the Royal Commission but at the price of little or no research.

(iii) Standing Bodies: Standing Royal Commission
The Standing Royal Commission may be set up by a special Act of Parliament, or through the Minister's prerogative or conventional powers. The method of setting up such a Commission may not be very important unless with an extremely contentious issue it is felt that an Act needs to be passed to give special powers and that the terms of reference need to be debated in Parliament. The Coal Industry Commission of 1919 was in fact an *ad hoc* Commission and not a Standing Commission and its mandate was thoroughly discussed in

Parliament as class and industry interests were deeply involved. There would perhaps be little point in appointing a Standing Commission in such a manner, as their terms of reference are usually very general indeed to allow great flexibility. As an example, the recent (1978) Standing Commission on Energy and the Environment's terms of reference are defined as: 'to advise on the inter-action between energy policy and the environment'.

These Commissions could have a very similar structure and the same procedural flexibility as the *ad hoc* Royal Commission, but they will be permanent bodies. Standing Commissions may be said to be truly advisory in nature rather than being a Commission of Inquiry into very specific problems. There will be elements of inquiry and investigation within the procedure following the lines of the Royal Commission.

A Standing Commission must be a part-time body of 'outsiders', and therefore subject to the criticisms advanced earlier. Care needs to be taken, however, that it does not become part of the departmental machinery dominated by Civil Servants. It does of course provide the opportunity for good working relationships to develop with various departments. As a permanent body it gives a continuity to inquiry and research and allows constant review of policy. It is possible that it would be able to oversee and advise on the implementation of its recommendations and to review the results. It may be entrusted to keep abreast of current technological (and political) developments at home and abroad and debate them from an impartial standpoint. The Commission would benefit from not being expected to come up with a definitive answer to a particular problem but would be allowed to adopt a more 'fluid' approach. Depending on its membership, the fact that it is 'outside' government should help to ensure its public acceptability.

A Standing Commission on energy could provide a much improved, higher level on-going debate on energy policy alternatives. Presumably the Commission would receive and call for evidence from alternative technology experts on a continuous basis and this may minimise the conflict that could occur at a public local inquiry.

Such a Standing Commission may not be geared to dealing with an 'urgent' energy issue, but (a) the urgency or otherwise of investments needs demonstration; and (b) single investments are better dealt with by LPIs or PICs while 'commitments' to programmes, existing or planned, should be dealt with by a Commission.

As with the *ad hoc* Commission, a Standing Commission has the ability to be a very public body: it can take evidence from anybody, make broadcasts, send out questionnaires (these are usually factual rather than opinion questionnaires) and send members on formal and informal visits. Membership is a problem and so is the role it could play in any debate on energy policy. Such a Standing Commission may need time to make its reputation and be trusted. It also needs to show it has some power and this must be shown in the fact that Parliament debates its reports and the Minister produces an answer to the report in some written, published or broadcast form.

(iv) Semi-standing Bodies: the Parliamentary Select Committee
There are Select Committees of the House of Lords and Select Committees of the House of Commons. They are appointed by their respective Houses and are therefore under the complete control of the House. Select Committee procedure appears to be much less flexible and public than that of a Commission.

Select Committee weakness derives from the fact that it is staffed totally by MPs, who have many other commitments and who are accountable to their respective parties. It is not always clear that they act as impartial assessors. The work of the Select Committee ceases when the Parliamentary Session comes to an end and it does not automatically continue at the beginning of the next session. Parliament must once more give its permission for the Select Committee to be set up. Hearings are usually held in the House and take a quasi-judicial form. The public are admitted to the hearings at the discretion of the Select Committee.

Select Committees are given power to send for people and documents but are not generally expected to be given the power to instigate their own research. It is unusual for a Select Committee to employ more than a single adviser. This is not to say that the work of a Select Committee is not thorough and its recommendations not challenging, but it is a comparatively inflexible procedure which is not very well publicised.

Experience suggests that the Government is very careful about giving too much attention to such a Committee of MPs, although it would give attention to MPs as individual representatives of pressure groups. It is highly doubtful that the Government would ever give a Select Committee a nationally contentious problem to deal with. (In August 1978 the Select Committee on Science and Technology called for CDFR to be investigated by that Committee, on the grounds that there had been a considerable period of inaction.)

In practice, three days of the Parliamentary year are normally given to the debate of Select Committee Reports. How far this is adequate depends on what issue is under debate. Select Committees do keep a selection of MPs informed about industry and technology issues and these MPs tend to have the power to ask 'awkward' questions in the House.

It would appear that Select Committees do not receive much money or co-operation. Part of the *Report of the Select Committee on Science and Technology, 1972–73* states:

> There is an obvious contradiction between the terms of reference given to Select Committees by the House, and the use of the Government of their power to control the majority in the House to limit the full execution of these same terms of reference.[4]

This behaviour would seem to reinforce the fact that a Select Committee may come into conflict with government by demanding more information than government is willing to give. Prime Ministers may be reluctant to strengthen the position of a Select Committee and consequently would prefer them to be told only what MPs are told in the Commons at Question Time. Disagreement continues in government about the delegation of power to Select Committees: in January 1978 the then Secretary of State for Energy, Tony Benn, is reported as having asked Michael Foot, the Leader of the House of Commons, to set up a Select Committee for Energy, saying that he would be prepared to give it all the documents it needed; but not many appear to share Mr Benn's enthusiasm for giving more power to the backbencher.[5]

We conclude that, while any contribution a Select Committee can make to the energy debate is welcome, it is not *alone* the appropriate body to investigate major energy investments or wider energy policy. We do however think that a Select Committee should make energy policy its concern.

(v) The Parliamentary Private Bill

The Private Bill procedure has been utilised in the past by large public companies who need special powers granted to them by Parliament. Complications and power struggles that would arise from trying to regulate nationally contentious proposals into Private Bill form would be considerable. Most importantly, a power struggle of some kind could develop between the Committee debating the Bill (if it reached a second reading) and the House of Commons in general. At present, Committees are totally dependent on the House for their power, including the power to discuss issues widely pertaining to the Bill. By giving wide debating power to a Committee, the House would conceivably lessen its own power of decision. This is of course if the Bill ever gets as far as Committee debate. For a Bill to get through a couple of readings would depend on the composition of the House and at least a 'neutral' Government. The Government has at the introduction of the Bill to make some stand and if the Bill is to be debated at all this stand must tend towards tacit or otherwise approval for its clauses.

The timing of debate would be important because of attendance of MPs. If there is any significant delay in proceedings the Bill must be postponed until next session and so on. Upon what information would MPs rely in order to make up their minds and reach a decision on the Bill? Presumably lobbies and pressure groups would be active. Information may, therefore, be biased and not all interested persons could reach the MPs. Furthermore, MPs may become embroiled in dilemmas of their own, constituency versus national interest, and personal opinion versus party, even though, in theory, there is a free vote. Private Bill procedure is too restrictive with regard to time to allow for the full debate based on research and thorough information that is warranted by energy issues. The Private Bill is a one-off event. The Bill is the result of the debate. It is not an on-going, discursive, advisory procedure. Many important issues could never be included in a Private Bill format.

This procedure is costly as well as rigid. Time is consumed by paying attention to fine procedural details rather than debate. All participants would find it expensive and difficult to understand without some legal advice. The drafting of petitions for and against the Bill would need careful consideration and the use of some legal aid. Objectors may only be heard on the grounds of their petition and their *locus standi* is also challenged on the points in their petition. It is possible for the House to give a discretionary right to a specially constituted committee on the Bill as to the hearing of petitions, but the total set-up is still prohibitive to the 'local' individual who does not wish to attach himself to any particular pressure group.

Committees are not allowed to hear evidence other than that which is tendered by the parties entitled to be heard without the express authority of the House. In other words they cannot instigate their own research, or call witnesses from a wider field of expertise than is offered by the participants, etc. Research would be quite impossible anyway due to the rigid timetabling that must be adhered to.

On the positive side the Committee may be given power to adjourn from place to place and its hearings may be carried on in public, but the hearings are very judicial in character. Cross-examination is confined to matters laid down in the petitions except when it is sought to discredit a witness. Hence if the original petition against the Bill confined itself to only a few clauses the objectors would find it very difficult to widen their case at a later date if they so decided. Presumably the Committee would if it deemed possible refer back to the House for a decision, an instruction or special permission.

Therefore, to clarify the points already raised, it would be fruitful to compare the Private Bill process, on a few issues, with that of the local public inquiry.

First, whereas an outline of the proposal may be sufficient for a public local inquiry the potential development in the form of a Private Bill would probably have to contain much more information. This is an advantageous aspect of the Private Bill.

Private Bill procedure could certainly subject the proposed investment to rigorous Parliamentary scrutiny in Committee and in debate in both Houses. Unfortunately, in so doing, the debate is removed from the locality for which the development is intended. This drawback is exacerbated by the fact that the procedure whereby objections may be lodged by members of the public against the Private Bill is far more complicated than that of the Public Local Inquiry. As previously stated, the *locus standi* of persons wishing to make objections to Private Bills has to be proved before the petition is allowed to be brought before the relevant Committee. Admittedly, it is likely that in cases where public concern has been aroused that *locus standi* will be awarded on very wide grounds. On the other hand, in practice, the Public Local Inquiry will afford a hearing to any party who wishes to speak without the same amount of pre-inquiry bureaucratic procedure. To conclude, the Private Bill, unlike the Public Local Inquiry, does not permit a debate open to comprehensive public participation and is therefore not a viable candidate for deciding nuclear and related investments on this one important issue alone.

(vi) The Ombudsman

The Ombudsman, or Parliamentary Commissioner for Administration, was provided for only in 1967. He is still, therefore, very much investigating the flexibility of the Act and the width of his powers. At present, the Select Committee which reviews his progress has proved to have a more radical concept of the role of the Parliamentary Commissioner than he has himself.

Originally it was felt that an Ombudsman would be a threat to the MP's role as the resort of the public in matters of maladministration, but instead he has been promoted as the champion of the backbench MP. The Ombudsman reinforces an MP's power to question government departments. His is primarily an investigative role which is set in motion by an alleged act of maladministration.

The Parliamentary Commissioner receives many hundreds of requests for help from the public. These requests are filtered by MPs and the acceptable cases given to the Ombudsman. His investigations must, therefore, proceed apace and be relatively inexpensive. Would it be fair to divert his attention from the 'small' matters of the individual citizen to an on-going energy debate or major single issue such as CDFR? This is not his purpose. If we were to instigate a special nuclear or energy Ombudsman the investigative process should give way to an advisory process. The limited provision the present Parliamentary Commissioner has for undertaking his investigation would have to be improved upon to an extent that in effect we should end up with something looking like a Royal Commission.

It must be remembered that the Parliamentary Commissioner for Administration can only act *after some sort of decision has been made* which will or already has adversely affected someone. This is self-evidently unsatisfactory with regard to energy issues. Usually the Ombudsman is not supposed to investigate a case which could have been dealt with in any other way, i.e. in a court of law, tribunal, etc., although in practice the Ombudsman can use his discretion in these matters.

It is evident that the Parliamentary Commissioner for Administration is not intended to provide a viable alternative to the public local inquiry. His role is to investigate alleged maladministration after the inquiry closes. The Ombudsman's examination is carried out fairly privately. The inquiry process is in two stages. Stage one involves liaison between the Parliamentary Commissioner's staff and the Department accused of maladministration. The Permanent Secretary of that Department is asked to comment and also undertakes his own investigation into the relevant representatives' actions. Stage two of the process consists of the examination of papers by the Ombudsman's staff and informal talks with the complainant(s) and those accused of maladministration. On this basis the Ombudsman will prepare a report and submit it to the MP through whom the complaint was made. In special cases, for example, where the Department refuses to institute a remedy for injustice, the report could be sent to Parliament.

It is not feasible, therefore, to suggest that the Ombudsman is a substitute for a local inquiry. His duty is to act as a check on inquiry procedure. No one at a public local inquiry stands accused of misconduct. The public inquiry is an investigative procedure but it is the merits of an application for planning permission which are under review, not the conduct of the intended developer.

On the whole, the Ombudsman is kept at a distance from the public, and his actions receive little publicity because his investigations are semi-secret. His existence tends to strengthen Parliamentary control. Parliamentary Commissioners have been chosen from the higher Civil Service; their staff are civil servants. The Office's privileged position of investigating private departmental papers means that the public is not able to assess the quality of the investigatory side of his work. The overall lack of clarity of the flexibility of his powers of investigation into the many facets of maladministration is an overwhelming drawback, and we conclude that the Ombudsman has no role to play in the decision-making procedure we are concerned with.

(vii) The Planning Inquiry Commission
The PIC, unlike some of the other administrative processes looked at, has a definite setting. It is part of the Town and Country Planning procedure. It is bound by certain statutory provisions but, as with the Act which set up the Parliamentary Commissioner for Administration, not a great deal is said about the actual procedural flexibility of the body. The Commission on the Third London Airport (the Roskill Commission) of 1968–71 is often cited as what a PIC might look like, but in fact it need bear little resemblance. It need not instigate the undertaking of a cost-benefit study, environmental impact analysis or other study, but it could do any of these. This power of independent research and investigation is vitally important, as we see below.

One point is clear, before a PIC can be set up *an actual application for planning permission* must have been tendered. This need not be a drawback if the plan is not urgent but in theory *there should be a fresh PIC for every new application*. We have already explained why, in our view, a *standing* body is needed for the review of energy policy. That is, energy policy cannot be properly debated by a succession of PICs investigating the 'components' of energy policy – a single fast reactor, a new coalmine and so on. There would be a lack of continuity if repeated PICs occurred. A Standing Energy Commission for energy policy is required, with appropriate *ad hoc* site inquiries of a local nature for 'localised' investments. But for planning applications which have a national impact a PIC is desirable. Note too, that while LPI's *can* make suggestions for alternative sites, a PIC can *explicitly* concern itself with an evaluation of alternative sites, as the Roskill Commission did. It is important that a PIC should in no way lead to restrictions being placed on the Standing Commission because of any conclusions it may reach. As far as CDFR is concerned, one benefit of a PIC over a Standing Commission will be that of impact on a particular issue. The press will report it, people will be prepared to take part in it. In short, it will produce a temporary burst of energy and enthusiasm on a *particular* issue. The Standing Commission will be much slower in general and perhaps not as newsworthy. But eventually, when it does have a particular piece of advice to give government or it is given a particular problem to debate by government, it could well have the same impact as a PIC while possessing the benefit of on-going rather than short-term research. In any event, if CDFR comes forward as a planning application only a PIC or LPI could investigate it under existing planning law.

The PIC has value in the fact that it draws the procedure of the 'Royal Commission' stage and the 'local inquiry' stage close together. In other words, under the Act, if Stage 1 is carried out then Stage 2 is almost automatically carried out. The Standing Commission and the local planning inquiry on the other hand are seen as two different separate processes, unless the Standing Commission is given power to investigate land planning policy for energy installations as well as general energy policy. The new Standing Commission on Energy and the Environment appears to have this potential. It will be asked to lay down broad planning guidelines but it will not be asked to look into specific planning applications.

A PIC would sit full-time on a specific problem whereas a Standing Commission is usually a part-time body. The Town and Country Planning Act 1971 is rather vague about the type of remuneration members of a PIC would receive and the amount of staff that should serve it. Furthermore membership of a PIC is limited. Although this could cut down on dissension it also cuts down the amount of expertise and representativeness, etc., of the panel. Theories have been promoted that the PIC could consist of a representative of a local area where the development is to take place, a planning inquiry inspector and other experts; but such a panel may be hard to 'balance' with an upper limit of five members, including an impartial chairman.

The PIC is mandated to hear the applicant, the local planning authority and legitimate appellants, i.e. those with a *statutory* right to be heard under the Act. Hopefully a PIC would be allowed discretion in practice to hear evidence from a wider sphere. Traditionally, national

Commissions may call in evidence from anyone: here precedent has already been set.

The final report of the PIC process, like that of the local public inquiry, would go to the Minister for his decision. Therefore it could not automatically be debated by Parliament. To enable such a debate to take place the original planning application would, like British Nuclear Fuels Limited's proposal, have to be made into a Special Development Order.

Nevertheless, it is feasible that any document produced in Stage 1 of a PIC could be debated by Parliament if some special arrangement was made. The difficulty here would be to 'feed back' into PIC procedure the 'results' of the debate. Certainly, the process would flow more easily if the general conclusions of Stage 1 were immediately put before a Local Public inquiry as foundations for continued discussion rather than be subject to an 'intermediate' Parliamentary debate.

We conclude that the PIC has a definite role to play in energy planning. In so far as energy investments can be deemed to have 'national' aspects they should be debated within the PIC framework. If, however, this national aspect is construed as including a commitment to an energy *programme* then it is proper that the investment be debated by a Commission which, in our view, should be a Standing Commission.

As noted above, PICs have independent powers of research and investigation. This may be contrasted with the powers possessed by a local inquiry inspector. The latter are extremely limited. The Inspector does not have access to a research team, although he can, and often does, request simple tests to be undertaken, or documentation brought forth. The potential research powers of a PIC are significantly greater. In the case of THORP, which raised matters requiring detailed analyses of engineering data, economic assessment, international law, radiation hazards and so on, these powers therefore become all the more important. It is totally unclear that such wide-ranging issues, of considerable complexity, can be adequately assessed by one inspector with one or more assessors. Additionally, the power to appoint five commissioners for a PIC would permit a much wider scope of expertise to be brought to bear on the issues. As an example, none of the assessors nor the Inspector at WPI had economic expertise, so that, despite having accepted the relevance of economic assessments, no effective debate or cross-examination took place on this issue (see Chapter 7).

If the PIC is so evidently the relevant institution for planning applications with this national dimension, why then has it never been invoked? The reasons appear to be bound up with a mixture of fear of the untried and fear of repeating an experiment – the Roskill Commission Inquiry into the location of London's Third Airport – which generated so much controversy. It seems worth dwelling on these motives since the suggestion here is that they have unduly prejudiced the development of planning inquiry procedure and have led directly to *ad hoc* modifications of local inquiry procedure, such as WPI, which, for reasons to be stated, we regard as distinctly retrogressive.

First, the PIC is technically reserved, if it is ever brought into force, for issues of national importance, probably with alternative site options. Indeed, Mr Justice Parker himself drew attention to this in suggesting that criticism of the WPI based on the idea that it should have taken the format of a PIC 'appeared to me of little merit' (Parker, 1978, para. 15.13). BNFL had made a specific application to site THORP at Windscale precisely because of the existing facilities located there. They made no application for an alternative site. Two comments are in order. In the first place, given the value of a PIC in this local/national context it seems evident that a PIC would have a valuable role to play because of its investigatory powers. The change suggested is slight and not by any means major. Secondly, although Mr Justice Parker posed as his second question the issue of whether reprocessing should be located at Windscale, it is far from evident that the proceedings would not have benefited from having alternative sites suggested and evaluated. Essentially, the arguments for keeping THORP at Windscale arise from the evident economies of concentration and the much-disputed argument that the hazard arising from THORP, which is additional to any existing hazard, still, when cumulated, came well within international limits of acceptability (for this debate see Chapter 7). The argument against would be that, if THORP is to be built at all, an alternative site would 'spread' risks in such a way as to lower them in absolute terms for the two sites compared to the situation in which THORP is sited at Windscale.

Second, a PIC would quite possibly involve greater expense than a 'modified' LPI. This in itself cannot be an objection, however, unless it can be shown that the extra cost, if it occurs, is less than the extra benefits accruing from the superior investigatory powers of a PIC. In his *Report*, Mr Justice Parker remarks on his own investigatory powers at the WPI, and those of his assessors to the effect that 'I regarded it as my duty to, and accordingly did, take steps to investigate any matter which appeared to me, or to either of my assessors, to require investigation whether or not it had been raised by any of the parties' (Parker, 1978, para. 15.12). None the less, if we consider the investigations that did take place, they did *not* include any independent assessment of the engineering feasibility of THORP nor any detailed economic assessment. In short, those investigations called for by the Inspector, valuable as they were, can hardly be described as constituting an assessment of the major issues relevant to the recommendation he was required to make. Given the sensitivity surrounding the Windscale plant, it is quite arguable that the benefits of an independent research assessment in terms of the efficiency criteria we have discussed would, alone, have outweighed any costs.

Third, it seems clear that, if a PIC was to be established, it would resemble in format the Roskill Commission of 1969–71. That Commission had seven commissioners and a research team of its own. It first considered a long list of possible sites for the location of London's (then proposed) Third Airport and progressively 'screened' them to reduce the final list to four. Local inquiries were then heard at the four sites (reversing the procedure one might expect from a PIC) and a major public hearing on the relative virtues of all four sites then took place in London in 1970. Those proceedings were, by and large, also adversarial (although no oaths were taken) and lasted some five months. The Commission produced a majority report (six of the seven members) and a minority dissenting report (one member). The majority recommendation was debated in Parliament and the vote there was *against* the recommendation of the majority of the Commission that the airport be sited at Cublington, north of London. Instead, the Parliamentary recommendation was that it be sited at one of the other four sites, Foulness, a coastal location in Essex to the east of London. Notably, Foulness had only been included in the Commission's short list to illustrate the superiority of the inland sites – it was not ever on the list of the Commission's four most desired sites.

Now, apart from the element of farce in appointing a major and expert Commission to make a recommendation only to have that recommendation rejected and one of the originally *least* desired sites chosen instead, what in the proceedings of the Roskill Commission should have occasioned criticism? For, if our argument is correct, it is this criticism that underlies part of the fear of introducing a PIC which, it is thought, would look all too much like another Roskill Commission. This may have been reinforced by the fact that even the Foulness site was abandoned as traffic forecasts proved to be exaggerated – although, interestingly, the pressure is now on again to introduce expanded capacity in the London region with a fourth runway proposal for Heathrow and the steady if almost surreptitious expansion of traffic at Stansted Airport in Essex, ironically the airport which, before Roskill, had been chosen for expansion and the protests about which had generated the Roskill inquiry itself! None the less, the fact that Parliament chose to reject a Commission's findings is hardly an indictment of the Commission's *procedure* (or even its recommendation – there are no oracular properties about Parliamentary votes). Rather the explanation seems to lie elsewhere entirely.

What caused the political fuss was that the Roskill Commission charged its research team with the preparation of a cost-benefit study of the four sites. The researchers embraced that technique, in its conventional form, to the full in the sense that money valuations of everything were attempted. This included human life and even historic churches, although the money valuation of the latter was quickly dropped in the light of criticism. What then happened was that the actual proceedings of the Commission were largely dictated by this cost-benefit study. Indeed, apart from planning matters, it seems fair to say the procedure involved those who opposed the cost-benefit study results not only offering theoretical and empirical criticism but actually having to present their own counter-studies to come up with a different result. Without going into detail here, cost-benefit analysis has always been a controversial subject, with its apparent concern to place money values on all issues (human life, the annoyance due to noise, landscape, the saving of travel time, and so on). In practice, few studies go as far as the Roskill study went and many economists would not espouse the extreme rigidity of approach which characterised that study (see Pearce, 1971, and Dasgupta and Pearce, 1972). Yet, to this day, the Roskill cost-benefit study is cited as the example of how economists should go about such studies – see for example Lovins (1977), who seems unaware that cost-benefit can have immense value without extending itself to the valuation of many intangible items. Cost-benefit studies need not embrace the Roskill methodology. Some would embrace that methodology and others would not: there is no *single* way of carrying out such studies.

If we then ally the political events that followed the Roskill recommendations with the fact that those recommendations were based on the cost-benefit study we see that the Roskill Commission become a political issue in itself. What was on trial was economics versus planning assessments, and cost-benefit analysis in particular with all its emotive attributes relating to the money valuation of the quality of life and, indeed, life itself. Could it be, then, that what opponents of PICs fear is a repeat of the political in-fighting that followed Roskill? It seems more than likely. Yet, if so, it reflects an unfortunate and erroneous assessment of what is likely to happen. In the first place, criticism of a *methodology* embraced by that Commission is not at all the same thing as criticism of the *procedure* of the Commission. The argument here is that, whatever one's view of the nature of cost-benefit analysis, *it is irrelevant to the format of an inquiry procedure.* Indeed, that inquiry procedure was itself exactly what was needed for the issue in question. For no one can seriously suggest that every conceivable aspect of the siting of an airport was not discussed thoroughly, even if the form of the discussion was dictated by the cost-benefit model.

Passing a little further into the realms of fantasy, could it be that opposition to a PIC is based on the fear that a PIC would embrace the same cost-benefit methodology? If so, the fear again seems irrational, not least because cost-benefit analyses have varied considerably in their treatment of issues since Roskill and also because wider research documents such as environment impact statements have come into use, albeit, on a limited scale in the UK.

There are 'legal' objections to the PIC as it is currently designed under planning law. But, its use in the case of the WPI would have greatly strengthened the investigatory powers necessary to deal with an issue of such wide-ranging implications.

AN OVERVIEW OF INSTITUTIONS

In terms of the energy debate, we noted two general requirements. The first is for a procedure to deal with localised investments which cannot be held to possess attributes which require them to be debated on the 'national agenda'. In principle, this requirement is met by the local public inquiry. Single investments may never even reach the local inquiry level. Most do not. Those that do generally reflect local or other pressures to secure a debate. Having accepted that the LPI is a 'proper' context for localised investments, however, in no way implies that local inquiries as they are currently carried out are 'efficient'. Indeed, in the next chapter we suggest a number of modifications to their procedure which, in our view, would enhance their value. In some cases we suggest that their *functions* should be changed to reflect an informational role that is currently not mandatory for such inquiries. This emphasis on information reflects the general emphasis on the informational require-ments for democracy repeated throughout this report.

The second requirement is for an on-going process to review, assess, investigate and advise not just on new programmes of energy investment but *also* on existing and past programmes. The latter aspect is essential if only because of the lessons that can be learned from '*ex post*' evaluations of programmes, such as their forecast and actual cost, whether safety standards have been achieved and so on. The emphasis on new programmes will be self-evident. They must be debated somewhere and will in general be debated in Parliament. We have stressed however that the very complexity and wide-ranging number of energy decisions to be made in the next thirty years or so cannot be left to a small body of persons whose traditional role it is, and must remain, to take ultimate decisions. The more 'outside' advice they have from sensibly constituted bodies, the better their role of vigilante and of securing public accountability will be. We have suggested that planning issues be left in the planning framework, with that framework suitably modified. These modifications consist mainly of changes in the rules of procedure and function of local inquiries and the explicit introduction of planning inquiry commissions to deal with single investments with multi-site features and/or national attributes. To deal with new programmes, however, we suggest a national and standing commission charged with the continuous review of energy policy in all its aspects. To the nature of this commission we turn in the next chapter.

It was stated that such a commission, in whatever form, should also look at existing programmes and past programmes. The difference between these two is perhaps not significant, any existing programme being the result of past decisions. The need for '*ex post*' or 'on-going' review is readily apparent when one considers the errors that have been made in the field of energy policy in the past. We do not dwell on them here, but the combination of private government and inadequate outside assessment of programme decisions largely accounts for the massive cost overrun of the advanced gas-cooled reactor programme (Henderson, 1977). Very detailed analysis of the AGR decision can be found in Burn (1978). These assessments alone should be sufficient to ensure that a review body be established to supplement the work of the Parliamentary Public Accounts Committee and such bodies as the Select Committees of the Houses of Parliament.

One other advantage of such a commission is that it could do much to remove the politics from local planning inquiries. It is arguable that what has happened in the growth of planning institutions is that planning frameworks have had to absorb an increasing political dimension as alternative value systems are debated in a context quite unsuited to such a debate, if only because planning law largely precludes the questioning of accepted national policy. Thus, taking the national policy element out of the planning framework (largely, since it is difficult to see how it would be removed readily from the planning inquiry commission deliberations) could restore the strengths of the planning system while simultaneously ensuring that the national dimension is debated.

From the preceding analysis it is evident that only a standing Royal Commission or a Standing Commission with full investigative powers, the power to secure information on demand, and without the disadvantages of existing Royal Commissions or standing commissions, will suffice. In short, the members need to be full-time persons, and there must be guarantees built in to prevent the Commission becoming an arm of government committed to Government policy. At the same time, it must be subject to measures of control which would allay fears that it will 'supplant' the ultimate right of decision of Ministers and MPs. We consider these attributes in more detail in the next chapter.

Finally, it will be noted that we have dealt entirely (with the exception of Select Committees) with extra-Parliamentary institutions. This is deliberate. It may well be that reorganisation is required within government and its agencies to improve energy decision-making. Those interested will find such a critique as far as nuclear energy is concerned in Burn (1978). To consider the internal structure of government and its agencies, however, takes us well beyond the remit set to ourselves. It is for others to consider whether the changes to the planning and energy advice procedure suggested here themselves entail changes inside government establishments.

NOTES

1. Much of this section is based on Cartwright (1975).
2. As one Royal Commissioner put it to us, Royal Commissions 'consist of people who have enough time to give to outside work and are willing to do this for little or no pay. Hence they are recruited from the ranks of the rich, the retired and the academic community, with far too little representation from the business world (including the non-academic professions and manual workers). Nor can they put in enough time.'
3. See Cartwright (1975).
4. Flood and Grove-White (1976).
5. Fenton (1978).

REFERENCES

Cartwright, T. J. (1975)   *Royal Commissions and Departmental Committees in Britain* (Hodder & Stoughton, London).

Flood, M., and Grove-White, R. (1976)   *Nuclear Prospects* (Friends of the Earth in association with the Council for the Protection of Rural England and the National Council for Civil Liberties, London).

Fenton, J. (1978)   'Power to the Corridor', *New Statesman* vol. 95, no. 2444.

Burn, D. (1978)   *Nuclear Power and the Energy Crisis* (Macmillan, London).

Henderson, P. (1977)   'Two British Errors', *Oxford Economic Papers* July.

Parker (1978)   *The Windscale Inquiry: Report* (HMSO, London).

Pearce, D. W. (1971)   *Cost-Benefit Analysis* (Macmillan, London).

Dasgupta, A. K., Pearce, D. W. (1972)   *Cost-Benefit Analysis: Theory and Practice* (Macmillan, London).

Lovins, A. (1977)   'Cost-Risk-Benefit Assessments in Energy Policy', *George Washington Law Review*, vol. 45, no. 5, August 1977.

Lidderdale, Sir D. (ed) (1976)   *Erskine May's Parliamentary Practice*, 19th edition (Butterworth, London).

5 Existing Advisory Institutions II

THE PUBLIC LOCAL INQUIRY

The main function of a Public Local Inquiry into a planning issue is to gather information to 'inform the Minister's mind', enabling him to reach a decision on a particular planning proposal. This is for cases where the inquiry has been 'called in' by the Minister.

Theoretically at least, the debate is confined to the one site issue and strictly planning matters are discussed. The applicant begins the proceedings, the objectors lodge their complaints, the applicant replies and the Inspector decides. In practice, public local inquiries have been very flexible. There may be more than one Inspector advised by a number of assessors. Furthermore, alternative sites may be discussed, albeit illicitly in the view of some. Additionally, the WPI launched into a discussion of national energy policy. This was a major departure, as a local inquiry generally does not discuss 'need' in the national sense.

The rights and roles of the applicant, local authority, statutory objectors and even their counsel are reasonably clear but the *de facto* rights of the non-statutory objectors are not. Basically a statutory objector is someone who is materially affected by the development or, for an application referred to the Secretary of State, anyone who has made representations to the Secretary of State within a specified time from the announcement of the development application. The WPI gave the non-statutory objector the same *de facto* rights as other participants and even allowed these objectors the same right of summary at the close of the debate as the applicant. After the debate, statutory and non-statutory objectors have some legal rights, although, at the time of writing, some doubts exist on this issue. Both statutory and third-party objectors may attempt recourse to the High Court but such an appeal could only be made *on a point of law*: their success would be highly dubious in the planning inquiry context. The objector must prove an injury to some legal right possessed by him. Moreover it is unlikely that the proceedings of the Inspector could be challenged in a court of law if he complied with the procedural rules. Further, his recommendations cannot be found to be right or wrong.

Participants are generally only entitled to costs if they are successful statutory objectors who appeal successfully against the unreasonable behaviour of opponents in compulsory purchase order cases or planning appeals. Participants need not be on equal footing since the appellant body may be better endowed financially than many objectors, or vice versa.

In a public local inquiry the Inspector is totally reliant on the evidence put forward by the interested parties. He has limited powers to initiate his own research. Hence the adversarial process may be defended in this case, as it would seem to be the only way of securing much evidence in the absence of a Freedom of Information Act.

Apart from having wide discretion over proceedings during the debate the Inspector also has complete control over what he admits as facts in his final report subject to the guidelines on adequate reporting. This report may be seen in three stages: (1) findings of fact, (2) inferences from the facts, and (3) recommendations. His report goes directly to the Minister in referred cases. It is not viewed in England by participants until the Minister publishes it with his decision. Furthermore the Minister is bound to make his decision on the evidence put forward in the report; if fresh evidence comes to light participants have the right to demand a reopening of the inquiry. In practice there is no way of knowing how the Minister reaches his decision. The reasons he publishes may not be those that directly influenced him. Complications are increased by the distinction laid down between the receiving of expert assistance on the evaluation of technical evidence given at the inquiry and experts' opinion on matters of fact. Participants may demand a reopening of the inquiry if the Minister has taken into account any new evidence or any new issue of fact (other than Government policy).

LPIs are of comparatively short duration compared to other administrative processes. The WPI took 100 days – it could have taken much longer. Nevertheless a great deal of planning can be carried on beforehand by the parties in preparation for the inquiry, in briefing of counsel, preparation of submissions, documents and research. It is, therefore, important that the public are notified in plenty of time that an inquiry is to take place. Again the non-statutory objector is in a much worse position than those with a legal entitlement to be heard. Officially, notice of a public inquiry should be given to statutory parties not less than 42 days before the inquiry is due to begin. For referred inquiries the Rule 6 statement by the Secretary of State, laying down the remit of the inquiry, is also served generally not less than 28 days before the inquiry. The third-party objectors must wait for notice of the inquiry in the local press, which is put in by the local planning authority when they see fit. Obviously all parties may be aware of the setting up of the inquiry well in advance of when they are entitled to hear, but the fact is that non-statutory objectors have much more of a 'detective' role to play than the statutory participant.

WPI, like the normal public local inquiry, could well have by-passed a less vigilant Parliament. Parliament has no right to debate an Inspector's report. The March 1978 Parliamentary debate set a new precedent but in the wrong context. With the demand of Parliament to take part in the Windscale debate the public local inquiry has been elevated to national level. This has produced a complexity where none need have existed. By not being able to release the report to Parliament *before* deciding on it, the Secretary of State for the Environment had to revert to a procedure of refusing the application (although he agreed in principle), having a debate in Parliament, and then having to introduce a Special Development Order to give BNFL permission. The Order can itself be debated (and was in the THORP case, albeit briefly). This situation seems unlikely to change with a PIC but Parliament could interact with an Energy Policy Commission, making its views known, debating reports and perhaps having MP representation on the Commission.

RIGHTS AND DUTIES OF PARTICIPANTS IN THE REFERRED LPI

A local public inquiry may be set up under the Town and Country Planning Act 1971 (hereafter called the 1971 Act). Such inquiries are constituted for various planning issues. The WPI was set up under Section 35 of the 1971 Act, which enables the Secretary of State for the Environment to call in a planning application for review. More commonly, local public inquiries are formed under Section 36 of the 1971 Act, which allows a developer to appeal against a refusal of planning permission by the local authority. It is important to note here that, in the main, if the local planning authority is 'minded to approve' a planning application no public inquiry will take place. It would appear that the applicant for planning permission has a right to develop the land if the local authority so permits and in this situation the public has no right to challenge their so doing. Hence public participation in planning control has been regarded as a privilege in the past, but current general opinion deems it to be a *de facto* right. Planning law must be seen to be flexible enough to cope with such changing social trends.

The LPI is governed by the Tribunals and Inquiries Act 1971 (TI Act 1971) and the Town and Country Planning (Inquiries Procedure) Rules 1974. Where a public inquiry is set up for some other purpose and is not conducted in accordance with a set of procedural rules, but has a quasi-judicial purpose, then it must in any event obey rules of natural justice. The rules of natural justice are 'minimum standards of fair decision-making imposed by the common law' (Young, 1978). Procedural rules may be seen as a parallel requirement to the tenets of natural justice, and can be considered to exhort the inquiry to improve upon the tenets by requiring a wider definition of 'fairness' than the rules of natural justice would warrant.

The two rules of natural justice are as follows:

(i) *nemo judex in causa sua* (nobody is to be judge in his own cause);
(ii) *audi alteram partem* (hear the other side).

The second rule may be interpreted as requiring that all parties to an inquiry must be given a fair hearing. These precepts have many ramifications and frequently when the Minister's decision on a planning issue is appealed against in the High Court the challenge will be made on the grounds of non-compliance with natural justice.

An LPI, although organised on quasi-judicial lines, is *not* a court of law; no one is to be found 'wrong' or 'right'. The inquiry is a quasi-judicial process in an administrative framework, designed to bring all relevant evidence to light in order to apprise the Secretary of State of the fullest information possible for him to make a reasonable decision on the issue. It may be seen therefore that procedural rules are not only designed to provide a fair hearing for the parties concerned but also to ensure that the Minister, who must review the facts 'at a distance', is adequately advised.

To recap, the public local inquiry is a flexible planning procedure which incorporates the tenets of natural justice and rules of procedure. It is important to determine the rights and duties accruing to each participant in the light of the role prescribed for the local planning inquiry by the relevant acts and statutory rules.

(a) The Inspector

The procedural rules allow the Inspector to use his discretion in the running of the inquiry except in instances where a specific rule has been made. It is within his discretionary power to permit non-statutory objectors (see section (*f*) below) to call evidence and cross-examine persons giving evidence. Therefore the Inspector has a right to limit these privileges. For example, Rule 10 (3) of the procedural rules states: 'The applicant, the local planning authority and the Section 29 parties shall be entitled to call evidence and cross-examine persons giving evidence, but any other person appearing at the inquiry may do so only to the extent permitted by the appointed person.'

The admission of evidence at the inquiry is also at the discretion of the Inspector and power of veto may be used. In cases where protection of the national interest is involved evidence is inadmissable. The Inspector is entitled to consider any written representations or statements he receives prior to an inquiry from any person as long as they are disclosed at the inquiry in order that the parties may be given an opportunity to comment.

As shown, the procedural rules confer discretionary powers on the Inspector which provide him with the opportunity of flexible action. Unfortunately it is difficult to determine the limits of an Inspector's discretionary rights, although it may be said that the use of these powers should not lead to an abuse of the rules of natural justice. The conflict that may arise between the Inspector's discretionary powers and the rules of natural justice may pose problems for the High Court. This point is discussed again later.

The Inspector is mandated to report in writing to the Secretary of State for the Environment (or, more recently, for road inquiries, the Lord Chancellor's Office). This report must include his findings of fact and recommendations, if any, or his reason for not making recommendations. The procedural rules say nothing about the format of the report and nothing more exact about its content than mentioned above. The guidelines on report writing issued by the Department of the Environment recommend a format and style of composition for Inspectors for local inquiries held under Section 36 of the 1971 Act. These are also issued to Inspectors conducting local inquiries under other sections of the 1971 Act. The guidelines were drawn up to encourage Inspectors to make their reports as concise, fair and informative as possible so that the Minister is not forced to differ from his Inspector on a finding of fact or to consider new evidence which would give interested parties an opportunity to press for a reopening of the inquiry. The notes stress that the report should satisfy the participants to the inquiry that their evidence has been adequately and fairly recorded. As paragraph B. 2.19 of the notes of guidance for Inspectors (*B2 – Reporting*) states, 'The Inspector's job is to marshal the arguments in a logical and effective manner, each point being made once only, so that the strongest possible case is made for each party, regardless of his own view or the merits of those arguments.'

The guidelines do not impose any mandatory duty on the Inspector and it is within his power to disregard the suggestions therein. Conversely he must comply with the *letter* and *spirit* of the procedural rules which the guidelines may be judged to have expanded and structured.

(b) The Assessor
An LPI may take place with or without an assessor(s). The function of an assessor is to evaluate specialist evidence and to advise the Inspector on the weight he should give it in coming to his conclusions. The incorporation of the assessor's views of the evidence in the Inspector's report is a matter for the appointed person's discretion. Once the Inspector has included the assessor's views, he accepts total responsibility for them. Where he has been dependent on the assessor's advice the guidelines for Inspectors state that this should be clearly identified in the report. In cases where a decision hinges on specialist issues the assessor is enabled to write his opinion, together with any supporting arguments, as an appendix to the report. The procedural rules do not refer to the role of the assessor, who therefore has no *statutory* duties imposed upon him.

The public expect the assessor to be a 'neutral' adviser – that is to have no professional connection with any of the parties involved in the inquiry. Wraith and Lamb (1971) mention a case where the assessor at an inquiry into the proposed erection of a fertiliser plant had previously had contact with the applicants and the planning authority in his role as head of the Ministry of Housing and Local Government's Alkali Inspectorate. He had been consulted about the technicalities of the proposed plant. As a consequence the objectors felt that there had been a breach of natural justice because a member of the Alkali Inspectorate, which had approved the proposal in advance, was now acting as assessor at the planning inquiry. Although this case was not brought to court it has since been agreed that where the Alkali Inspectorate has had prior dealings with an inquiry participant an assessor should be appointed from outside the Department where possible. It may be taken as a general principle that the Department of the Environment will strive to appoint assessors who have no prior connections with the case. If this is not possible and the appropriate inspectorate has already been consulted and the appointment of an officer from that inspectorate is unavoidable, then the assessor selected should be of equal or senior rank to the person who had previously been involved with the case. The role of the assessor is fundamentally different from that of the Inspector. The Inspector is not obliged to take the assessor's opinion into account on major issues. The assessor's views are only important to him where specialist knowledge can be of service. Consequently the assessor's role is limited and subordinate to that of the Inspector who is in charge of the inquiry. Any legal challenge the parties to an inquiry wish to make about non-compliance with statutory duties will rebound on the Inspector and the Secretary of State for the Environment, not on the assessor, because, although all parties must comply with the relevant Acts, the assessor has no legal role in that he is not mentioned in the Act.

(c) The Local Planning Authority
In Section 36 cases of the 1971 Act, the Local Planning Authority will take part in the public inquiry as a defender of their decision to oppose the proposed development. That is to say the authority would have refused to give the applicant planning permission (or may have failed to give a decision), thereby causing the developer to appeal against their action, thus resulting in the setting up of an inquiry. Where the application for planning permission has been called in by the Secretary of State for the Environment under Section 35 of the 1971 Act the planning authority may appear in the capacity of a supporter or objector to the proposed development. Suffice it to say that, whatever the case, the local planning authority as the responsible body for the overall development of the locality is in a special position at the local public inquiry. The procedural rules lay down certain standards which this authority must meet to enable a fair public inquiry to take place. It is therefore necessary to discuss some of the more important duties imposed on the local planning authority.

The rules provide for the Secretary of State to make the local planning authority responsible for publicising the Inquiry. Furthermore the authority is bound by the rules to serve on the applicant, the Section 29 parties (statutory objectors) and the Secretary of State a written statement(s) of the submission(s) it intends to put forward at the Inquiry. If it decides to put documents in evidence it has a duty to send out a list of the same with its statement(s) of submission(s). These documents must be made accessible to the public and where practicable the local planning authority must provide the public with the opportunity to make copies. A written statement of submission requested from the applicant by the Minister may be included in these documents. During the Inquiry, subject to the discretion of the Inspector, the local planning authority may be enabled to alter or add to its submissions and list of documents if it will facilitate the determination of questions in controversy between the parties.

Non-compliance with any of the above-mentioned duties would be seen as a breach of the procedural rules and perhaps of natural justice as these statutory functions facilitate the accumulation of information by the public with regard to the planning application and the local planning authority's views on it.

(d) The Secretary of State for the Environment

The Secretary of State or relevant Minister and his department bear the responsibility for making the final decision on planning applications for development which have gone to local public inquiries. As previously stated, a referred inquiry of this nature is only set up when contentious issues arise. Frequently, important planning decisions contain an element of national policy and it is therefore reasonable to assume that in a number of cases the Minister will not accept his Inspector's findings because of other factors he must take into account. In cases where national policy will be an important element in making the final decision the Secretary of State may issue a policy statement for the benefit of the Inspector and Inquiry participants. Where such a statement is not forthcoming, Rule 6 of the procedural rules may provide an alternative indication of the type of evidence the Minister will regard as relevant. Rule 6 mandates the Secretary of State to compose a written statement of his reasons for having an application referred to him. He notes the points which seem to him likely to be relevant 'to his consideration of the application' (Rule 6, procedural rules). These points may be few and purposely general to enable the Inquiry to undertake far-ranging discussion. On the other hand they could clearly indicate the type of evidence that the Minister expects to receive.

It is important that the Secretary of State's decision should be seen to be 'fair and reasonable' despite the fact that he may refuse to adhere to his Inspector's recommendations. The procedural rules lay down certain criteria for the reopening of an inquiry in cases where the Minister has acted outside the limits of his statutory power by rejecting the Inspector's recommendations for no valid reason. Hence some of the parties to a public local inquiry are entitled to make representations for the reopening of that inquiry in two instances. First, where the Minister differs from the Inspector on a new issue of fact; and second, where the Minister takes fresh evidence into account. Unfortunately these two criteria have raised problems of interpretation in the High Court. Nevertheless the Minister is bound to give reasons in writing for his decision which must be seen to be fair and rational.

(e) The Applicant for Planning Permission
Requests for planning permission may be granted or refused without any inquiry ensuing. But since the application may result in an inquiry, the law imposes upon the applicant for planning permission quite a number of duties. In return he is entitled to demand a public inquiry, be notified of the date of the inquiry, secure the 'Rule 6' statement of the Secretary of State in referred cases, and the statement of submission with accompanying list of documents from the local planning authority. Often, the Secretary of State may require the applicant to serve a statement of submissions and list of documents on himself, the local planning authority and the Section 29 parties.

During the Inquiry the applicant has the right to make the first statement and also has the right of final reply. He possesses the right to call evidence and cross-examine witnesses. The proponent may also alter or add to his statement of submissions during the Inquiry at the Inspector's discretion. On the whole, the procedural rules aim to ensure that the applicant has very few more rights at the inquiry than the statutory objector. As noted above he may be required to make a statement of submissions in advance of the inquiry to the benefit of the objectors who are not under the same obligation. Of course it is possible for the Inspector to request of the objectors a pre-Inquiry statement.

(f) The Objectors: Section 29 Parties and 'Other Interested Persons'
The difference between a Section 29 objector and what are termed 'other interested persons' by the procedural rules is the fact that the former has either availed himself of the right to make representations to the local planning authority under Section 29 of the 1971 Act or is a person with a material interest in the land proposed for development, being its owner or an agricultural tenant. Under Section 29 of the 1971 Act the objector becomes a statutory participant in a referred Inquiry. Other participants appear merely at the discretion of the Inspector and are non-statutory objectors. In practice this distinction makes very little difference to the privileges afforded to both at the Inquiry. Where this distinction could

matter is if appeal to the High Court is considered after the Inquiry.

Prior to the public debate all objectors are entitled to take copies of the 'Rule 6' statement of the Minister, the statement of submissions by the local planning authority and similarly the statement of submissions by the applicant.

Section 29 parties enjoy the same claim as the applicant and the local planning authority to the calling of evidence and cross-examination of witnesses. Non-statutory objectors 'may do so only to the extent permitted by the appointed person [the Inspector]' (Rule 10 (3), procedural rules).

A more fundamental difference in the entitlements of statutory and non-statutory objectors is that of making a claim for the reopening of the Inquiry if an action by the Minister so warrants. A non-statutory objector is denied this right. The prevalent view is that if an error has been committed by the Minister and substantial prejudice thereby issues to the applicant or statutory objector, these latter will make representations to the High Court if they have sufficient standing. If they do not object it is implied that they are satisfied with the decision and their right not to act should not be superseded by any action on the part of those objectors with a less valid interest. This is a very fine distinction between the role that the statutory and non-statutory objector plays in the LPI. Such procedural vagaries may cause problems on appeal to the High Court by non-statutory objectors.

As discussion of the procedural rules may have indicated, the nature of the public local inquiry is complicated. It exists within an administrative and legal framework, is subject to quasi-judicial procedure and may undergo judicial review. The vagaries of the procedural rules, especially with regard to the discretionary powers of the Inspector and the rights and privileges of the non-statutory objector, may cause problems for the High Court. Each judge may interpret individual participants' rights in a different way. It is appropriate to consider some of the major problems of judicial review of local public inquiry procedure.

JUDICIAL REVIEW OF THE LPI

First, if any party to a LPI wishes to make an appeal to the High Court in order to obtain a quashing of the Minister's decision there are only two grounds of challenge possible and both relate to errors of law. Challenge may be made on the grounds that the Minister's order or action is beyond the powers of the Act. This is the challenge of 'substantive *ultra vires*'. That is to say that the Secretary of State has gone beyond the limitations set down by statute for his actions. The second ground of challenge is if the interests of the applicant have been *substantially prejudiced* by a failure to comply with any of the relevant requirements. 'Relevant requirements' in this case means any requirements of the 1971 Act, the 1971 Tribunals and Inquiries Act and the procedural rules. This is the challenge of 'procedural *ultra vires*'. The challenge must be made no later than six weeks from the date on which the order is confirmed or the action taken.

The grounds for appeal are restricted for a number of reasons, one of the main ones often advanced being that a public inquiry is a costly procedure and the decision arising therefrom should not be quashed for minor defects in the procedure which, in any event, are ruled out by the requirement to demonstrate 'substantial prejudice'. Furthermore, the Minister has complete control over planning policy and so the High Court should not have wide powers to question his decisions. The balance between the Minister's, the local planning authorities', the applicants' and the objectors' rights could be tipped in favour of those appealing against the Minister's decision if wider grounds of challenge were available. This could cause a slowing-down of the administrative process as well as undermining the Secretary of State's elected power.

Currently the High Court is able to quash the Minister's decision for a variety of reasons within the scope of the grounds of challenge. Mention will be made of some of the most frequent reasons for quashing a planning decision. If the High Court determines that the Minister took irrelevant, or failed to take relevant, considerations into account in reaching his decision it is likely to be quashed. This would also be the case if the Secretary of State formed his decision upon no evidence; i.e. if the Inspector's report did not contain enough information for the Minister to come to a rational decision.

Rule 13 of the procedural rules exhorts the Minister to notify his decision and the reasons for it in writing to certain of the inquiry participants. If these reasons are found to be unintelligible they would provide grounds for the quashing of the Minister's decision.

As indicated previously in this chapter the most widely used ground of challenge concerns the scope of the Minister's powers in respect of substantive or procedural *ultra vires*.

Having given a brief review of the grounds of challenge open to the participants to the public inquiry it is necessary to determine their standing in a court of law.

THE PARTICIPANTS' STANDING IN A COURT OF LAW

(a) The Appellant

A person appealing to the High Court against a Minister's decision is often called the 'appellant'. The appellant could be either the applicant for the proposed development, a Section 29 party or a non-statutory objector. The first hurdle the appellant must overcome is to satisfy the court of his *locus standi* (right to be heard). It is at this juncture that inquiry participants discover the legal limitations to their rights and privileges. The area of greatest contention is that of the *locus standi* of non-statutory objectors. Persons with a legal interest in the land in question automatically possess *locus standi*. The Turner case in 1973 (see below) has enlarged the number of persons entitled to *locus standi* by giving other Section 29 parties similar rights, but there remains some confusion on this issue. As Section 245 of the 1971 Act states, a 'person aggrieved' by an order or action of the Minister may take an application to the High Court to quash such order or action. Unfortunately judges in the High Court have differed as to the interpretation of 'person aggrieved'.

In the case of *Buxton* v. *Minister of Housing and Local Government* which was heard in 1960 a 'person aggrieved' was judged to be a person who had suffered some legal wrong; the decision must have wrongfully deprived him of something.

The applicant and Section 29 parties have *statutory rights* set down in the procedural rules but other objectors (and, for that matter, supporters) appearing at the public inquiry do so at the Inspector's discretion and are thereby accorded a *privilege*. The narrow definition of 'person aggrieved' evinced in the Buxton case meant that Major Buxton, who had appeared at the public inquiry to oppose an application to dig chalk on land to which he was a close neighbour, was denied statutory remedy on the grounds that no legal right possessed by him was infringed by the grant of planning permission.

Since then, the definition of 'person aggrieved' has been revised by the High Court, notably in *Turner and Another* v. *Secretary of State for the Environment and Another* (1973). Ackner J. declared in this case that any person who has attended and made representations at the inquiry should have the right to establish in the High Court that the decision is bad in law for some good reason. This definition would include non-statutory objectors who appear at the discretion of the Inspector. They must still show that they have suffered substantial prejudice by the

Minister's action (if the ground of challenge is in respect of procedural *ultra vires*), thus preventing frivolous appeals.

The Turner case indicates how the Courts reflect current trends in the public local inquiry process; whereby privileges are beginning to develop into rights. It is important though to remember that vagaries still persist where the post-inquiry rights of objectors are concerned. It is possible that Ackner J.'s ruling could be reversed in a subsequent case, so that possible inquiry participants should avail themselves of the rights designated to objectors under Section 29 of the 1971 Act.

It could be argued that all those who take part in a public local inquiry by giving evidence should automatically receive *locus standi* if High Court action comes about. Arguably, the Turner case actually produced this situation. The Buxton case indicated the previous reluctance to give 'person aggrieved' a wide definition. In making the wider definition, the Turner case may have blurred the distinction between interested parties to a local inquiry and the general public. Any person giving evidence without being directly affected by the development, but who had an adverse opinion of it, could, on this interpretation, have recourse to the High Court to quash the Minister's decision. Others, however, feel that Ackner J.'s judgment is suspect (Purdue and Trice, 1974). At the time of the Buxton case, for example, if an error of law was committed then the applicant or the statutory objector would challenge the decision. Therefore there was no need to extend the right of challenge to those with a less valid interest. Furthermore if the number of people allowed recourse to the High Court was increased the planning process could be seriously slowed down.

The current position appears to be that those who take the trouble to prepare evidence, call witnesses and use up their monetary resources are responsible persons with a legitimate claim to post-inquiry rights equivalent to those of the applicant for planning permission.

(b) The Inspector
The High Court faces a number of difficulties if challenge is made on the action of the inquiry Inspector. There may have been non-compliance with the procedural rules, in which case the action the High Court takes should be clear. For example if the Inspector has denied the applicant the right to cross-examine witnesses this would be so. The issue is complicated when an Inspector's discretionary powers are involved. To take the previous example of the right of cross-examination the Inspector may deny this right to a non-statutory objector with regard to any witnesses. Upon consulting the procedural rules it will be found that

the Inspector in exercising this discretion has acted within the letter of the rules. Nevertheless it is interesting to note that this discretionary power was recently tested in a case not subject to the procedural rules of the Town and Country Planning Act, *Nicholson* v. *Secretary of State for Energy and Another* (1977), where it was held that an objector who could be a 'person aggrieved' had the right to cross-examine witnesses who had given evidence contrary to his case. 'Although the inspector had a discretion, that discretion could not be exercised to exclude the rights of non-statutory objectors if that exclusion would be contrary to the rules of natural justice' (*Times Law Report*, 5 August 1977). A person who complains of a breach of the rules of natural justice has a heavy burden where he is unable to show a breach of the procedural rules. Therefore it may be noted that the Court will insist that the discretion has been exercised within the law and especially in compliance with the tenets of natural justice.

The High Court's interpretation of the Inspector's role is very important. The Inspector is the Secretary of State's representative at the public inquiry. It is his job to hear evidence, choose the salient facts and present them in a cogent form to the Minister. The High Court do not find it to be their function to rehear the evidence as presented to the Inspector. It will confine itself to reviewing legal flaws within the Inspector's report. There is no redress for errors of fact, only errors of law (but see Gibbs' case below). This means that the High Court will insist that the findings of fact comply with general legal principles including natural justice. Nevertheless, it is left to the Inspector to decide how much weight he attaches to each party's case.

With respect to the adequacy of the appointed person's report *W. H. Gibbs Ltd.* v. *the Secretary of State for the Environment and Another* (1973) is an interesting case. The public inquiry in this example was into a Compulsory Purchase Order. The objectors submitted to the Court that the Inspector's report was too confused to present their case properly to the Secretary of State. Consequently it was found by the Court that the report was so lacking in precision and intelligibility that it could not have been of any real value to the Minister. Furthermore, it appeared to the Court that an essential part of the objector's case was not put before the Secretary of State who, as a consequence, did not have the material before him to judge the case properly. His decision to confirm the CPO was duly quashed.

From the Nicholson and Gibbs cases it may be inferred that privileges afforded to inquiry participants under the discretionary powers of the Inspector may become 'rights' under the principles of natural justice which are upheld by the High Court where an Inspector invalidly exercises his discretionary powers. It must be remembered though that discretionary privileges are not in the same category as statutory rights.

(c) The Secretary of State
After the Inquiry the Minister bears the responsibility for reaching a decision on the evidence presented to him.

Usually the Secretary· of State will attempt to reach a balanced decision based on the evidence in the Inspector's report. He will try to balance the interests of the individual with those of the locality and the nation. Fundamental to this aim is a balanced report which will enable the Minister to see all sides of the case and how the Inspector came to make his recommendations.

If the Minister abuses his administrative power the High Court will provide a check. Non-compliance with the relevant planning Acts would result in the High Court quashing the decision. Non-compliance with the terms of any relevant Act would consist of the Secretary of State not carrying out duties mandated to him in the terms of that Act. Abuse of his power would occur if he misuses his discretionary rights or takes power upon himself which is not given to him by the Act. For example, if the Minister held consultation with one party to an inquiry without the knowledge of other parties this might constitute a breach of natural justice, particularly if prejudice could be shown, and possibly an abuse of his discretionary powers. Non-compliance with the procedural rules with regard to the receipt of fresh evidence and errors in the letter of decision are cited in cases brought before the High Court.

The Minister must demonstrate in his letter of decision that he is able to give adequate reasons on relevant considerations and that these reasons are within the power of the 1971 Act. Arguably, judges have shown themselves to be wary of quashing a Minister's decision on grounds of unreasonableness where it contains an element of national policy, especially where the Court is able to find some evidence to support the decision. The challenge of unreasonableness rests on either procedural or substantive *ultra vires*. For example, the Secretary of State may be found to have disregarded relevant evidence or, conversely, to have had regard to irrelevant evidence. This would be likely to prove that the Minister had made the decision without having adequate or proper evidence for so doing.

The letter of decision may indicate that the Minister has considered other evidence than that provided by his Inspector's report. The inquiry participants, not having been given an opportunity to comment on this evidence, may accuse the Secretary of State of non-compliance with the procedural rules. Thus the problem arises of whether the Secretary of State has the authority to discuss the political implications of the planning application with his advisers and other departments. If this discourse comprises detailed evidence of a local nature then it is likely that the Inquiry should be reopened in the interest of natural justice, but if the advice is of a general nature it may not be considered necessary to reopen the inquiry in order to satisfy these rules.

Where the Secretary of State has differed from his Inspector on a finding of fact and is thereby disposed to disagree with his recommendations, and has failed to allow parties to the inquiry to make representations to him concerning this he is in breach of the procedural rules. Difficulties arise for the High Court on the ascertainment of whether the Minister has actually differed from the Inspector on a fact or merely dissented with his opinion. In *Lord Luke of Pavenham* v. *the Minister of Housing and Local Government* (1967) it was held that by refusing Lord Luke planning permission against the recommendation of his Inspector the Minister was not differing on the findings of fact of his official but was expressing a difference of opinion on a planning matter – i.e. the emphasis to be placed on the national policy which precluded sporadic development in the countryside. Consequently the Court did not quash the Minister's decision, having no right to interfere with the way in which he carries out planning policy provided he has not committed an error of law.

To conclude, the scope of the High Court's power to intervene is not always clear where the rights of the Secretary of State over administrative action are concerned.

CORRECTIVE MEASURES

(a) Public Rights and the Local Planning Authority
It may be considered unfortunate that the public is often only made aware of the deliberations of the local planning authority concerning the proposed development when the plan is subject to a local inquiry. Only the bare outline of an application which is sent to the LPA for consideration is published in the local press. Nevertheless, open meetings are instituted by the LPA for discussion of large-scale development and people are entitled to attend and express their opinion. It is unlikely that alterations to the proposal could be put forward at this stage in order to negate or lessen objections to the plan

prior to a local inquiry because discussion at this stage will not be in depth. Only when a public local inquiry is instituted will cases be fully prepared by all those intending to participate. It could be argued that the planning process would be retarded by automatic public review every time a planning application was put forward but it is unlikely that people would want to be drawn into a discussion on every planning application. The LPA could make a distinction between proposals for development requiring public inspection at the application stage and those that do not. To entitle the interested public to participate at this earlier stage on contentious planning issues would be seen to be democratic and reasonable. Furthermore, participation at this stage need not preclude a local public inquiry if it became clear that certain contentious issues required deeper investigation.

(b) Rights of the Non-statutory Objector
As formerly referred to, the system of the local public inquiry is currently flexible enough to allow the non-statutory objector the same rights during and after the inquiry as those whose rights are clearly laid down in law. These 'rights' are bestowed on the non-statutory objector by the Inspector at the Inquiry and the fact that the High Court may accept the right to *locus standi* according to its interpretation of the law. As seen, the non-statutory objectors' 'rights' in a number of instances are either intimated by the procedural rules or are deemed to exist within the general tenets of natural justice. It is unlikely in the contemporary climate of opinion that these 'rights' will be eroded. Should this vague distinction between the rights of the statutory and non-statutory objector be left as it stands?

The apprehension may still exist that if legal rights are conferred on all who choose to join in a public inquiry then the High Court will be inundated with appeals. This is unlikely to occur for several reasons. First, any party who is willing to bear the cost of a High Court action will have a valid challenge to make. Second, if the inquiry procedure is carried out according to law then there can be no ground for complaint. Third, if the participants believe that they have had as fair a hearing as possible and have had their evidence adequately represented to the Minister it is impossible that they would feel it necessary to challenge the decision: An alternative view which may be advanced is that if the objectors consider that the Minister has made a 'wrong' decision they will still take steps to upset it.

Nevertheless, what is suggested here is that the present non-statutory objector be given the same legal right as the statutory objector to seek a *post-inquiry* review by the High Court in instances where one of the grounds of challenge can be shown.

(c) Information

From the foregoing examination of inquiry procedure it is evident that comprehensive information pertaining to the planning application should be made available to the public prior to and during the inquiry. At present the Secretary of State requires the planning authority and the applicant if he controls the land to post notices about the impending inquiry in newspapers in the vicinity and in conspicuous places near to the land in question. Where a planning application has a national dimension it would be advisable to advertise the inquiry in the national press.

The pre-inquiry meeting which may be held by the Inspector should become compulsory and potential inquiry participants required to attend. The secretary to the meeting should receive a list of the documents that each party to the inquiry intends to bring forward in evidence. In this manner the parties will be fully apprised of the type of information that will be introduced. Furthermore, to enable such exchange of information to take place presupposes that the Rule 6 statement of the Secretary of State is available to the meeting. Based on the points raised by the Rule 6 statement the pre-inquiry meeting should be able to discuss the likely scope of the inquiry so that everyone has a general idea of the form the inquiry will take and the type of issues likely to be important.

Either at the pre-inquiry meeting or within a specified time and place all parties should be obliged either by law or on the request of the Secretary of State to produce copies of the documents they intend to submit for consultation by the public. It appears important to stress that the participants should be compelled to bring forward any relevant information within their possession on the merits or demerits of the application. At present adverse information is only brought to light by the adversarial process which is probably less efficient at bringing the defects of a development to light. It may be argued that a developer will not unearth information harmful to his case. Thus the imposition of a penalty for withholding information might be necessary. Furthermore all significant government documents should be made available on request by the Inspector after discussion with the parties about the content of the Secretary of State's Rule 6 statement.

Therefore an efficient pre-inquiry meeting would not only determine inquiry procedure but also ensure the availability of comprehensive information to the parties who intend to become involved in the inquiry. Once in possession of complete information the participants would be able to prepare a competent and valuable case.

(d) Timing

Changes made relating to procedural matters must result in an examination of the timing of the public inquiry process. A comprehensive pre-inquiry meeting will require a set minimum of months, prior to the public inquiry, for digestion of important material and preparation of evidence. The preliminary meeting before the Windscale Public Inquiry was held on 17 May 1977, just under one month before the Inquiry began. As this was just a meeting to determine the programme of the forthcoming Inquiry it is reasonable to assume that this was adequate, but it does not seem unreasonable to suggest that a more comprehensive meeting be held about six months in advance of the public inquiry. This period of time would not only permit adequate preparation of evidence but would also give the objectors an opportunity to communicate with each other concerning the handling of different issues pertaining to the inquiry. In this way the possibility of repetition of evidence at the inquiry could be reduced.

Currently the Minister is under instruction to give not less than 42 days' notice of an inquiry. This may be adequate notice for a local issue such as an application to build a house or a shop but it is an inadequate intimation of notice for more important issues such as the WPI. Admittedly it can be left to the Secretary of State's discretion to give longer notice in more contentious cases. The problem recurs with regard to the Rule 6 statement which at present may be served to the participants up to twenty-eight days before the date of the inquiry. Again this is inadequate notice for the reasons previously mentioned.

(e) Post-inquiry Meeting

In Scotland, participants to the local public inquiry receive a draft of the Reporter's (Inspector's) findings of fact, upon which they are allowed to comment before the complete report also containing the appointed person's recommendations is sent to the Minister. Apparently this is a valuable practice which, it is reasonable to assume, cuts down on the need for judicial review. In England this habit has not yet been adopted because it might slow down the planning process. To present 'findings of fact' as a separate document for comment would suggest that the Inspector follows the Department of the Environment's guidelines on report writing. The format laid down here indicates that there should be a distinct section headed 'Findings of Fact' prefaced by the words 'I find the following facts . . . ' Presumably the sections which present the cases of the various participants would also be included in the draft 'findings of fact'.

Oral as well as written comment could be valuable on Part One of the Inspector's report. Thus a post-inquiry meeting chaired by the Inspector could debate suggested alterations in the findings of fact. Thereafter the Inspector would be entitled to use his discretion in composing the final draft. The need to have recourse to the High Court may be significantly reduced by this practice. Obviously a number of persons may disagree with the Inspector's final report but they would be unable to say that they had not received a fair hearing. Furthermore, a comprehensive report would serve the function of adequately informing the general public. It may therefore be concluded that the Inspector, not only in his report, but throughout the inquiry has at least a moral duty to ensure the ventilation of all issues for the information and education of the public.

Even after the close of an inquiry information will continue to be produced. Hence a post-inquiry meeting on issues of fact could take into account new factual information but would not be open to a debate on opinion. On the whole it is doubtful whether many new facts will be brought to light in the time between the close of the inquiry and the production of the draft findings of fact. Some will see a danger of the post-inquiry meeting escalating into a renewed public inquiry. As stated above this is not its purpose; this should be made clear. This does not dispel the problem of the Minister receiving new information independently of the inquiry.

These changes have their own risks, namely the possibility of creating a context in which the Inquiry may be reopened and the opportunity increased for procedural challenge. Clearly the matter is deserving of more study than has been possible here.

IMPLICATIONS FOR THE PLANNING INQUIRY COMMISSION

Up to date no specific procedural rules have been made for the running of Planning Inquiry Commissions. Certainly the Town and Country Planning (Inquiries Procedure) Rules 1974 do not apply to an inquiry which might be held by a PIC. Nevertheless undertakings have been given that such an inquiry would be conducted 'within the spirit' of the above-mentioned statutory rules. This of course means that the inquiry run by a PIC would not only have the same flexibility as a local public inquiry but also be subject to the same drawbacks that have been examined throughout this section.

One of the major drawbacks is the fact that the final report of a PIC could not automatically be debated by Parliament. The report, like that of a local public inquiry, goes to the Minister. Therefore to enable a debate in Parliament to take place the Minister must reject the planning application. It would appear that the proposed development must be introduced into Parliament as a Special Development Order, upon which a vote is taken.

This process seems to negate any rights participants in a local public inquiry (whether part of PIC procedure or not) may have had to raise legal objection to the form or content of the Inspector's report or the Minister's action thereon. In the normal course of events the Minister's decision on a planning application is open to judicial review. The debate on a motion to annul a Special Development Order by Parliament is a legislative, not an administrative act and is therefore not subject to judicial review.

REFERENCES

Foulkes, D. (1976)   *Introduction to Administrative Law* (4th edition, Butterworth, London).

de Smith, S. A. (1973)   *Judicial Review of Administrative Action* (3rd edition. Stevens & Son, London).

Wade, H. W. R. (1977)   *Administrative Law* (3rd edition, Clarendon Press, Oxford).

Wraith, R. E. and Lamb, G. B. (1971)   *Public Inquiries as an Instrument of Government* (Allen & Unwin, London).

Young, E. (1978)   *The Law of Planning in Scotland* (William Hodge & Co., Glasgow).

Town and Country Planning Act 1971   (HMSO, London).

Tribunals and Inquiries Act 1971   (HMSO, London).

Tribunals and Inquiries   Town and Country Planning (Inquiries Procedure) Rules 1974, SI 419.

B2: Reporting   Department of Environment Guidelines for Inspectors.

Windscale Local Inquiry   Note of an Informal Preliminary Meeting held at 10.30 a.m. on 17 May 1977 at the Civic Hall, Whitehaven, Cumbria.

Purdue, H., and Trice, J. (1974)   'Further Light on Judicial Definitions of "Aggrieved Persons" in Planning Law', *Journal of Planning and Environment Law*.

CASES

Buxton v. *Minister of Housing and Local Government* (1960), *All England Law Reports* 408 [Queen's Bench Division].

W. H. Gibbs Ltd. v. *Secretary of State for the Environment and Another* (1973) Queen's Bench Division.

Lord Luke of Pavenham v. *Minister of Housing and Local Government*, (1967), *All England Law Reports* 1066 [Court of Appeal].

Nicholson v. *Secretary of State for Energy and Another* (1977). *Times Law Report* 5 August 1977 [Queen's Bench Division].

Turner and Another v. *Secretary of State for the Environment and Another* (1973), 72 *LGR* 3801 [Queen's Bench Division]. See Foulkes (1976) above, p. 212.

6 The Windscale Inquiry: Technical Background

INTRODUCTION

One of the main reasons for the reprocessing of spent uranium oxide fuel is that of resource conservation. The fuel is termed 'spent' when it has been irradiated in a reactor for up to three years and is reaching the stage of diminishing efficiency in the production of heat through the fission of atoms. If retained in the reactor after this period heat is produced less efficiently and so the fuel is removed. 97 per cent of this irradiated fuel is uranium, up to 1 per cent is plutonium and 2–3 per cent is radioactive waste (Allday, 1977). As the fuel has been enriched – i.e. it has had uranium 235 added – it contains far more U235 after irradiation in a thermal reactor than is found in natural uranium. Reprocessing frees this uranium and plutonium from other actinides and fission products so that they may be reconverted into fuel for re-use in fast reactors and, conceivably, in thermal reactors. The effect of re-using the reprocessed fuel in a thermal reactor is to enable 30–40 per cent more power to be generated from the original material. However, such fuel can only used once. On the other hand recycling this material in a fast reactor would give an energy saving 50 times greater than that elicited from a thermal reactor (Allday, 1977). Hence the uranium and plutonium in spent oxide fuel is self-evidently a valuable source of energy if it can be recovered at a cost which permits the recycled U and Pu to have a value such that their use in thermal or fast reactors would provide power at a competitive price.[1]

Very few countries have adequate indigenous supplies of uranium. Most non-communist nations are therefore dependent on imports. The notable exceptions are the USA, Canada, Australia and South Africa. Reprocessing coupled with the development of a fast reactor programme would make a nation largely self-sufficient in electrical energy supplies and consequently not so subject to 'resource diplomacy'. Furthermore this process is considered by many to be essential for the maintenance of the option of a fast reactor programme, which would require the plutonium produced by reprocessing of spent fuel from thermal reactors and, to a lesser extent, the Pu 'bred' in fast reactors.

The possible shortage of cheap uranium supplies would also seem to indicate the wisdom of reprocessing if the nations are to make nuclear power their main source of energy. It would appear that a uranium shortage is not imminent. The date when supplies become limited would depend on the rate of growth of demand, development of a reprocessing industry, exploration and the discovery of new sources and the promotion of fast reactor programmes. None the less, a risk remains. Reprocessing (and the development of fast reactors) would lessen the risk of a shortage of energy.

Where the management of nuclear waste is concerned the UK nuclear industry has always assumed that reprocessing is an integral part of the nuclear fuel cycle. Consequently the 'throw-away' solution to spent fuel arisings has never been fully investigated, although extensive experience exists in, for example, Canada. On this 'solution' no spent fuel is reprocessed and it must be disposed of. By 1995 Great Britain will probably have 3300 tonnes of spent oxide fuel in storage from its Advanced Gas Cooled Reactors (AGRs) alone (Allday, 1977). Hence the alleged need for reprocessing.

THE NUCLEAR FUEL CYCLE

The Nuclear Fuel Cycle is 'The sequence of operations in which uranium is mined, fabricated into fuel, irradiated in a reactor and reprocessed to yield uranium and plutonium for re-use as fuel' (RCEP, 1976).

Reprocessing is not at present considered by all nations possessing nuclear power to be a necessary stage in the Nuclear Fuel Cycle. For example, Canada is investigating methods of retrievable storage of spent fuel elements. As yet the vast majority of countries which have acquired nuclear reactors do not command the facilities to engage in a complete nuclear fuel cycle. Therefore different stages of the fuel cycle may be carried out at great distances apart. For example, Japan obtains her supplies of enriched uranium from the USA but will have to transport her irradiated fuel elements to Great Britain and France for reprocessing.[2] Meanwhile, in Europe, Great Britain, France and West Germany are each attempting to construct a complete nuclear fuel cycle.

Three variations of the nuclear fuel cycle may be identified as suitable for Light Water Reactor fuel (Rose and Lester, 1978). First is the 'throw-away' cycle, where irradiated spent fuel is stored rather than reprocessed. The second would allow for the recovery by reprocessing of unused uranium but not plutonium. Recovered uranium would be fabricated into new fuel, thereby re-entering the fuel cycle, whereas plutonium would be treated as a waste product. The third and most widely accepted cycle is that which allows for the reacquisition of both the uranium and plutonium from the irradiated material by reprocessing. These fuels are thereby made available for fabrication into fuel for re-use in nuclear reactors.

Figures 6.1, 2 and 3 (Rose and Lester, 1978) describe the three variations of the nuclear fuel cycle.[3]

We may now look at each stage in the various cycles.

(a) Uranium Mining and Milling

This is the first operation in the nuclear fuel cycle. Mining is prosecuted by open-cast, underground and place-leaching methods (Surrey, 1973). The largest reserves of uranium ore are to be found in the USA, Australia, Canada and South Africa. After extraction the ore is milled by crushing and grinding the ore into fine particles which are leached with acid to yield a uranium solution which is further purified and converted to the oxide known as 'yellow-cake'. 'Yellow-cake' consists of about 90 per cent uranium oxide (U_3O_8). This material undergoes further crushing to form a powder. Waste products arising from milling, which are called 'tailings', are stored indefinitely.

(b) Conversion to Uranium Hexafluoride

In order to enrich the uranium for use in AGRs and LWRs the 'yellow-cake' must be converted into uranium hexafluoride (UF_6: hex). 'Hex', which is produced in Great Britain at BNFL's Springfield Works, vaporises easily, allowing enrichment to take place.

(c) Uranium Enrichment

Natural uranium contains the isotopes Uranium 238 (U238) and Uranium 235 (U235). U235 is the more fissile of the two isotopes and consequently sustains the chain reaction in a nuclear reactor. Fission involves the division of the U235 atom into two lighter atoms and spare neutrons, thereby releasing energy as heat. The extra neutrons collide with other fissionable U235 atoms. Continuation of this process is termed a 'chain reaction' whereby heat is constantly produced. Magnox

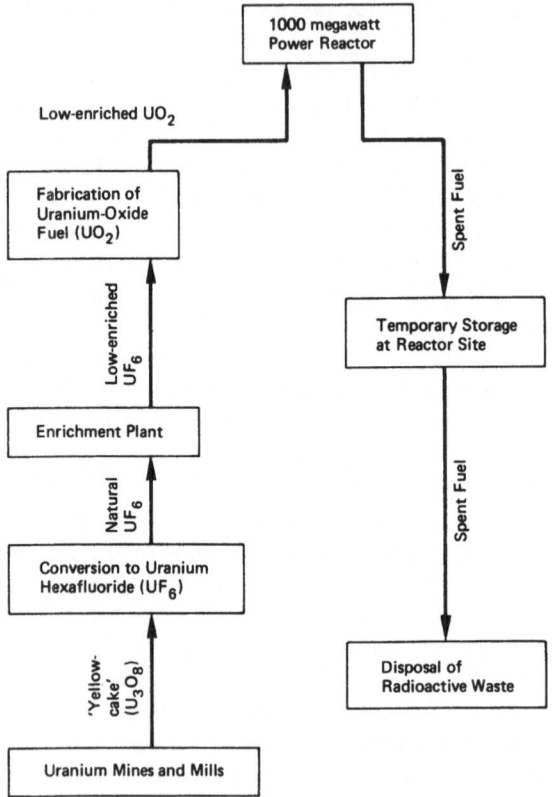

FIG. 6.1 The 'throw-away' cycle

fuel consists of natural uranium in which U238 is the predominant isotope. Only 0.7 per cent of this fuel contains U235. Although U238 is less fissile than U235 it can capture a proportion of the excess neutrons to form the more fissile plutonium, Pu 239. To promote a sustained 'chain reaction' in AGR and LWR fuel the proportion of the more fissile U235 in the uranium is increased from 0.7 per cent to 2–3 per cent. This augmentation in the proportion of U235 is called enrichment. Currently three enrichment methods are in use; gaseous diffusion (undertaken by BNFL at Capenhurst); gas centrifuge (also in use at Capenhurst); and the nozzle process. A further process, laser enrichment is under investigation.

The gaseous diffusion process depends on the difference in mass

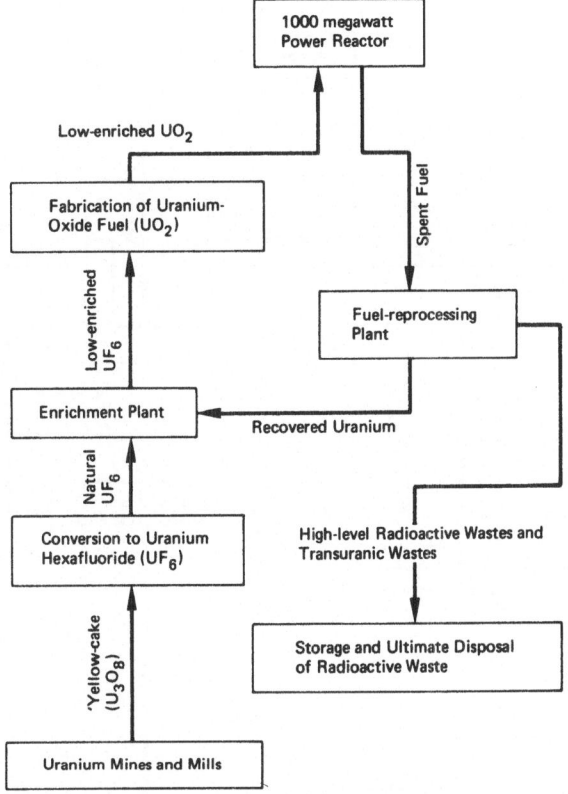

FIG. 6.2 The uranium recycle

between U238 and U235. These two uranium isotopes are separated from each other when the 'hex' molecules are pumped through a fine porous membrane. This pumping action uses a great amount of electricity. The lighter U235 molecules travel faster than their heavier counterparts. Hence the gas which manages to percolate through the membrane is slightly enriched and separated from the depleted gas which contains a slightly lower U235 concentration than before. The procedure is carried on until the optimum level of enrichment is achieved.

The gas centrifuge process consumes less electricity than the gaseous diffusion but it is subject to mechanical failure because of the highly stressed rotating equipment involved (Fox, 1977). It relies on a

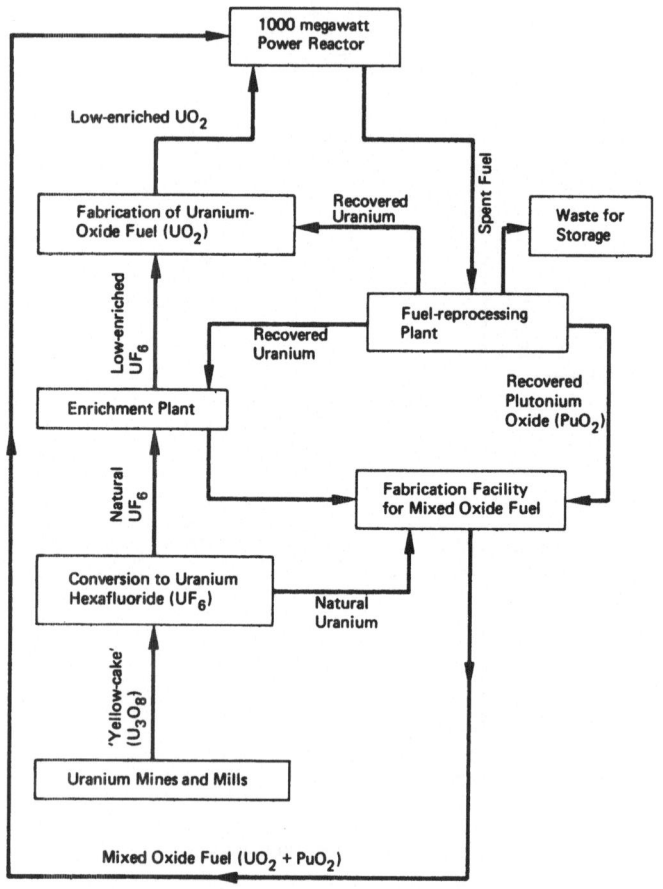

FIG. 6.3 The uranium and plutonium recycle with reprocessing

revolving centrifuge whereby the heavier U238 molecules in the hex drift to the outside and the lighter U235 molecules stay nearer the centre. This *modus operandi* allows for the separation of the gas into two streams. The process is repeated until the desired level of enrichment is procured.

Nozzle enrichment uses up large amounts of electricity, like the gaseous diffusion process. Here the 'hex' is mixed with hydrogen and together the gases are pumped through curved tubes which cause the lighter U235 molecules to be flung off at somewhat greater angles than the heavier U238 molecules, whereby the two streams may be separated. Repetition of this process ensures the right amount of enrichment.

It has been shown that a finely tuned laser is able to render a temporary electric charge to the U235 molecules which may then be 'hived off' from the U238 molecules. The laser process might require only one stage to complete enrichment.

(d) Fuel Fabrication

This is the fourth stage in the nuclear fuel cycle. Fuel may be fabricated from non-enriched uranium or enriched hex. The most widely used reactors in the world today are LWRs, of which the Pressurised Water Reactor (PWR) surpasses the Boiling Water Reactor (BWR) in popularity.[4] The enriched hex or the yellow-cake is converted into a uranium oxide powder which is formed into pellets and baked. The high density of these stable pellets assists the chain reaction, improves conductivity and helps to retain the gaseous fission products created by the reaction (Patterson, 1976). A batch of uranium oxide pellets are clad in either a stainless steel or zirconium alloy ('zircaloy') thin-walled tube to make a fuel pin. Before sealing, these pins are filled with an inert gas. Once sealed they are assembled into composite arrays.

(e) Reactor Refuelling

The fresh fuel is transported to the reactor where it is assembled in the core. Gas-cooled reactors may be reloaded while still running but those cooled with water must be shut down. The fuel elements are left in position for about three years, by which time they have ceased to function efficiently and are said to be 'spent.' Consequently they are removed from the reactor and placed in a cooling pond on site before shipment to a reprocessing plant or to an interim storage facility. The pond water is chemically treated in order to reduce the rate of corrosion and to remove any fission products which may have leaked out of the cladding. Storage on site allows for the heat produced by the decay of the irradiated fuel to dissipate and for fission product activity to be reduced.

(f) Transport

If reprocessing is to be undertaken the irradiated fuel elements are shipped to a reprocessing plant, such as that at Windscale. Transport of such radioactive material is governed by national and international regulations.[5] In Britain irradiated fuel is carried in heavy steel flasks which are lined with lead or depleted uranium. The spent material is placed under water within the flasks. Decay heat dissipation is also enhanced by fins on the external surface of the flasks.

(g) Reprocessing

In the case of Windscale, by the time THORP is constructed and operating the first batch of spent fuel will have been in wet storage for about five years, which will have further reduced its radioactivity (Warner, 1977). During this period some radioactive fission products will have decayed. Reduced heat and levels of radioactivity are benefits of lengthy cooling periods but prolonged wet storage may lead to the corrosion of fuel coverings (cladding), followed by leakage of fission products. The magnesium alloy (magnox) cladding of natural uranium fuel corrodes more rapidly than the stainless steel of AGR fuel or the zirconium alloy (zircaloy) surround of Light Water Reactor (LWR) fuel. It is accepted throughout the nuclear industry that spent magnox fuel must be reprocessed within about one year of wet storage but there is some uncertainty as to the length of time spent oxide fuels may be stored under water before reprocessing is undertaken. This uncertainty is greater for stainless steel-clad fuel than for zircaloy-clad fuel, on which there is fairly extensive USA experience.

The first stage of reprocessing spent oxide fuel requires the use of a 'Head-End' plant. The function of a 'Head-End' plant is to take uranium *oxide* fuel through the stages necessary to reduce it to a process liquid which can be introduced into the flow sheet of a plant designed to handle uranium *metal* fuel. At this plant fuel containers are clamped to the end of a series of 'caves' or 'hot-cells', which are heavily shielded rooms allowing for the handling of highly radioactive materials by remote control. Individual fuel elements, which are assemblies of separate fuel pins, are fed into the first cave (Patterson, 1976). Here a shear chops through the element, which may contain up to 100 fuel pins. The 1″–3″ pieces thus produced drop into a stainless steel container called the dissolver basket. This contains heated nitric acid which dissolves the fuel within the cladding. The pieces of stainless steel or zircaloy cladding (hulls) are cleaned and stored. The nitric acid stream containing uranium, plutonium, other actinides and fission products now flows into the main reprocessing plant. For magnox-clad fuel the removal of the cladding (decanning) is undertaken by purely mechanical means at the storage pond and the metallic fuel rods are then fed to the dissolver of the reprocessing plant.

THORP will use a combination of pulsed columns and mixer settlers in the solvent extraction process. Pulsed columns will be used in the first solvent extraction cycle and the later plutonium extraction stage because they minimise solvent degradation and accommodate the higher plutonium throughput without reducing the nuclear safety of the plant. Mixer-settlers will be used for the process which takes uranium out of the solvent stream and back into the nitric acid stream. Simplifying a very complicated procedure, the solvent TBP/OK extracts the uranium and plutonium from the other actinides and fission products which remain in the nitric acid solution. The uranium and plutonium are purified and separated into uranium and plutonium nitrates. After several further purification cycles the nitrates are converted into plutonium and uranium oxides. The uranium is returned to the BNFL Springfield plant for conversion into uranium hexafluoride for re-use in the nuclear fuel cycle. The plutonium recovered by reprocessing is stored at Windscale or converted into a plutonium-based fuel for use in the prototype fast reactor at Dounreay. Fuel made with this plutonium has also been used in the prototype AGR, the Steam Generating Heavy Water Reactor (SGHWR) and overseas LWRs (Allday, 1977). An additional stage may be introduced for the separation of plutonium and neptunium from the uranium. This will be followed by a further cycle to remove the neptunium from the plutonium product.

(h) Non-reprocessing or Limited Reprocessing Cycles
The preceding discussion of reprocessing has centred on the 'full' nuclear cycle whereby uranium and plutonium are recovered. The other two cycles differ in that in one uranium only is recovered and the plutonium is treated as a waste product.

In the case of the 'throw-away cycle' neither uranium nor plutonium is recovered. Effectively, therefore, they must be stored permanently.

Storage of fuel assemblies that have been removed from reactors poses technical problems. This method of dealing with spent fuel would involve the disposal in some form to the environment of *all* the plutonium and uranium remaining after irradiation. Hence a much larger volume of waste must be managed than that which is produced by reprocessing. Magnox fuel kept in wet storage is known to suffer from corrosion after one year. Discovery of corrosion and significant leakage of stainless steel or zircaloy-clad fuels after a prolonged period of storage would mean that some form of reprocessing would have to be undertaken in haste. This would create complications if no reprocessing facility for oxide fuel were available for immediate use. Protection from leakages occurring during long-term wet storage of irradiated fuel elements might be given by its 'bottling' or encapsulation. Such a process could be as technically complicated as reprocessing (Flowers, 1976). Corrosion may be reduced by dry storage (Conroy, 1978), a method which may involve the replacement of water by air or helium. In Britain, magnox-clad fuel is kept in dry storage at the Wylfa nuclear power station before being transferred to Windscale for wet storage. If dry storage becomes the acceptable procedure of dealing with spent fuel rather than reprocessing for long-term or permanent disposal, facilities will need to be redesigned to cope with increased quantities of material and different types of fuel (Colton, 1978). Currently research into methods of dry storage has reached an intermediate stage. Air-cooled vaults are employed for the storage of high-temperature gas-cooled graphite-moderated reactor fuel because the difficulties posed by construction, fuel handling and heat dissipation have been assumed to be sufficiently demonstrated. On the other hand, dry storage of Heavy Water Reactor (HWR) fuel has not yet been demonstrated, although Colton (1978) states that it is considered to be available. When development of various dry-storage procedures (air-convection vaults, air-conduction vaults, concrete canister containment, underground salt beds and caisson)[6] has been completed it may be possible to convert underground retrievable storage facilities into permanent disposal sites for unreprocessed, encapsulated fuel. In the meantime, spent fuels from HWRs, including the Canadian Deuterium Uranium Reactor (CANDU), which were originally intended for permanent post-irradiation storage, are still in wet storage as an interim measure.

Possible options for spent fuel management using various storage techniques are shown in Figure 6.4.

(i) Post Reprocessing

Where reprocessing does occur the recovered stocks of uranium and/or plutonium can be fed back into the fuel cycle at the enrichment or fuel fabrication stage. The reprocessed uranium from oxide fuel will still have a higher U235 content than natural uranium and consequently it may be transported straight away to the fuel fabrication plant for blending with depleted uranium to make the equivalent of natural uranium for reactors which can use that, such as magnox. It is unlikely to have any direct use in AGRs or water reactors, for which it would have to be re-enriched. The plutonium may either be stored or used to produce a mixed uranium/plutonium fuel for use in thermal or fast reactors.

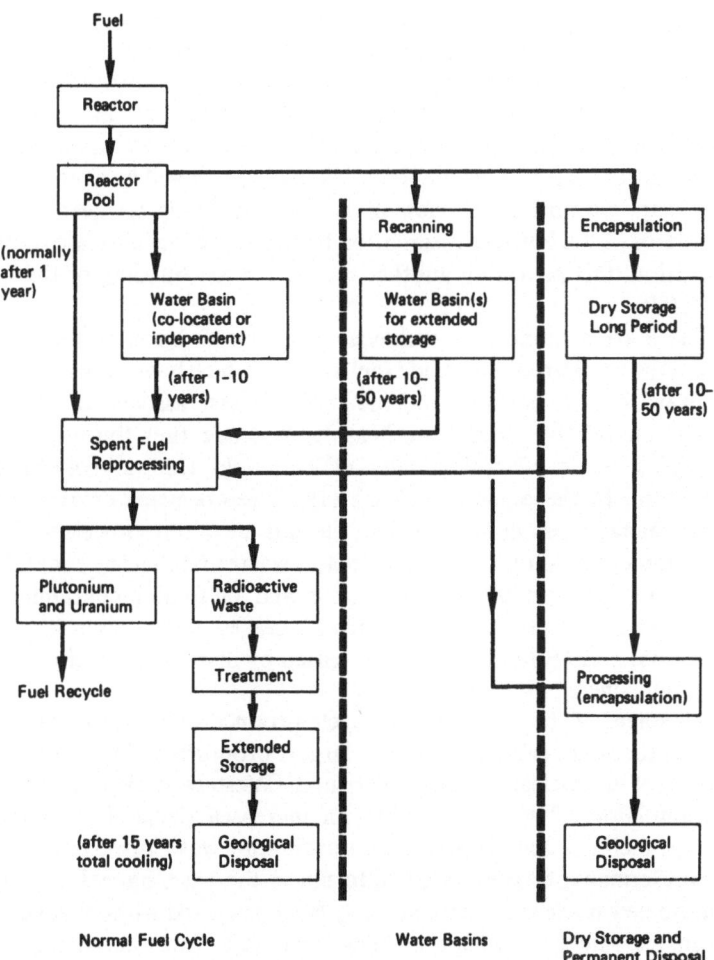

FIG. 6.4 Possible options for spent fuel management[7]

REPROCESSING AND WASTE MANAGEMENT

Reprocessing of over 20,000 tons of uranium metal fuel has been successfully completed at Windscale since 1952. Spent oxide fuel was reprocessed at Windscale between 1969 and 1973. By 1973 roughly 100 tonnes of spent oxide fuel had been reprocessed. Operations ceased after an accidental release of radioactivity in the B204 plant which, having been used between 1952 and 1964 for the processing of magnox fuel, had been modified to undertake the early stages of oxide fuel reprocessing. From 1964 magnox fuel reprocessing was carried on in the newer B205 plant. The disused B204 plant was converted into a 'Head-End' plant for spent oxide fuel in 1969. Consequently when further modifications have been completed the B204 plant will be over twenty-five years old and in the view of BNFL is unlikely to meet current standards of the Nuclear Installation Inspectorate for full-scale routine operation, this providing another reason for the building of the new THORP.

There are a number of technical problems to be mastered in the reprocessing of irradiated oxide fuel, most of which have been discussed in Franklin (1975). These include the problems associated with the higher cumulative output of heat from oxide fuel (burn-up) and increased levels of radioactivity which could lead to solvent degradation. In the opinion of Metz (1977) reprocessing creates more waste management difficulties than the storage of unreprocessed fuels. He comments that reprocessing increases waste materials by converting the spent fuel into acidic waste, hulls and plutonium-contaminated equipment. The process also involves releases of the gases Krypton 85(Kr-85) and Iodine 129(I-129), released by dissolution of the fuel to the atmosphere.

All stages of the Nuclear Fuel Cycle produce some waste materials but reprocessing creates by far the largest proportion. The continuing growth of the nuclear power industry in the Western world has led to the accumulation of waste materials, an increased demand for interim storage facilities and the need for co-ordinating waste disposal policies. West Germany, Sweden and California in the USA have determined that no new nuclear power plant may be constructed without adequate means of safe waste disposal being available. In Great Britain the Secretary of State for the Environment has instigated the setting up of a Radioactive Waste Management Advisory Committee, 'to advise him and the Secretaries of State for Scotland and Wales on major issues relating to the development and implementation of an overall policy for the management of civil radioactive wastes, including the waste management implications of nuclear policy, the design of nuclear systems, research and development, and including the environmental aspects of the handling and treatment of wastes' (*The Times*, 17 May 1978).

Effective waste management is important for the protection of future generations from radioactive emissions of long-lived fission products and actinides. Control of waste should be undertaken with the preservation of the natural environment as a priority.

WASTE MANAGEMENT

We can now discuss the management techniques applied to the waste products arising from reprocessing operations in Great Britain (unless otherwise stated).

(a) Gaseous Waste

Gaseous waste arises from 'hot cell' ventilation air, from vessel ventilation air and from the decanning and dissolution stage in reprocessing (Franklin, 1975). The various gases are treated chemically before they are released to the atmosphere and most of them are not expected to create environmental hazards. Krypton 85, a gas which is freed during the dissolution of the spent fuel, is not effectively removed by any of the treatments used to clean gases before they are discharged. Release of this fission product gas will increase with the operation of THORP. Mr Justice Parker recommended in his Report that a method of safe removal and retention of this gas should be investigated by BNFL and this recommendation has been accepted by the Secretary of State. BNFL have also accepted that a special Krypton 85 removal plant should be built if the technology is available. The Royal Commission on Environmental Pollution (RCEP, 1976) reported that the Atomic Energy Authority had developed a means of removing and storing Krypton 85. They report that it is proposed that this gas would be absorbed into active charcoal and then incorporated into a metal matrix. A similar problem applies to the release of the gas Iodine 129. An increased discharge of this gas to the environment could lead to increased concentrations of iodine in the human thyroid (Patterson, 1976). This would obviously need to be avoided. The technology for absorbing I-129 has been demonstrated in Germany and the USA and will be incorporated in THORP (Warner, 1977).

(b) Low-active Liquid Waste (LALW)

This is discharged from the Windscale works into the sea. Some LALW is the result of storage of medium-active liquid waste (MALW) which has since decayed. The major part of LALW consists of the con- taminated water from the irradiated fuel storage ponds, which is in the most part due to the fact that perforation of the magnox cladding through corrosion allows the water to dissolve small quantities of certain fission products out of the uranium bars. Further LALW is made up of effluent from the laundry, rainwater run-off and various process liquors. Before discharge to the environment LALW is treated to remove many of the radionuclides other than tritium. The main processes available are:

(i) 'distillation producing extremely pure distillates containing less than 0.001 % of the original activity';
(ii) 'floc precipitation processes which will produce purified liquid streams containing typically 1 % to 50 % of the original activity, depending upon the radionuclide present and the chemical conditions of the treatment';
(iii) 'delay of liquids in tanks to allow natural decay of relatively short-lived nuclides' (Franklin, 1975).

(c) Low-active Solid Waste (LASW)
Two kinds of LASW may be distinguished, that which is slightly contaminated by plutonium and that which is not. Franklin (1975) calculates that about 3–5 m³ of non-plutonium-contaminated waste will arise per tonne of reprocessed fuel. Waste items include tools, gloves, clothing and miscellaneous equipment. Material from the Windscale site which is not contaminated with plutonium is buried in trenches at Drigg. Plutonium-contaminated waste, of which 0.1 m³ arises per tonne of reprocessed fuel, is packaged and stored. Methods of plutonium retrieval, long-term storage and final disposal, probably to the Atlantic Ocean, are under investigation. The combustion of contaminated material coupled with the recovery of plutonium from ash is currently being used on a pilot scale at Windscale.

(d) Medium-active Liquid Waste (MALW)
MALW arises from the solvent extraction process, gas scrubbing, steam-stripping of high-active liquid waste and from the evaporation of the same (Franklin, 1975). About 60 m³ of MALW could be presumed to be produced for each tonne of fuel reprocessed.

MALW may either be treated by a process of evaporation and distillation or floc precipitation. The distillate is sufficiently free from radioactivity to be discharged to sea. The concentrate, which contains virtually all the activity fed to the evaporator, may be either mixed with HALW or stored to allow the short-lived nuclides to decay embodied in bitumen.

Floc precipitation creates a decontaminated waste product called a 'supernate' which contains roughly 5 per cent of the original activity. Sludge from the supernate is settled and de-watered for incorporation into bitumen. This bitumen product is packaged and stored for final disposal.

(e) Medium-active Solid Waste (MASW)

This may be defined to include the sludge produced by the treatment of MALW but otherwise MASW will include such items as fuel cladding, equipment and graphite sleeves resulting from the breakdown of AGR fuel elements prior to reprocessing (Clelland, 1977). Franklin (1975) describes fuel cladding as a High-active Solid Waste (HASW). He estimates that each cubic metre of zircaloy cladding from LWRs will be connected with roughly 10,000 curies (Ci) of fission product activity after one year of post-irradiation cooling. Currently only interim storage is in operation for MASW waste products; the cladding is kept in shielded silos. BNFL plan to clean the fuel hulls, compact and package them, or combine them with some other material such as metal, plastic or concrete prior to packaging. A cooling period (under water) of perhaps ten years is judged to be required before packaging may be undertaken.

Graphite sleeves will probably be stored in retrievable containers before final disposal. At present contaminated equipment is stored in silos but it is intended to decontaminate this waste and place it in containers so that it may be recovered when the possibility of further processing arises (BNFL hope to be able to treat this waste for final disposal by 1992).

Medium-active plutonium-contaminated waste includes gloves, plastics, tools and miscellaneous equipment. The limited amount arising at present can be packaged and disposed of into the deep ocean but a joint BNFL, UKAEA and Ministry of Defence development programme is under way to find alternative methods of containment (Clelland, 1977).

(f) High-active Liquid Waste (HALW)

It is estimated that about 5 m^3 of HALW results from each tonne of fuel that is reprocessed but this high burn-up oxide fuel liquid waste can be concentrated to roughly 0.25 m^3/t by evaporation (Franklin, 1975). This waste contains over 99 per cent of the activity of the fuel after one year of storage (Clelland, 1978). 650 m^3 of HALW concentrate, the acquired total of twenty-five years of British reprocessing operations, is at present stored on the Windscale site in tanks shielded by thick-walled concrete vaults (Clelland, 1978).

The most serious problems of containment and disposal are associated with HALW, which is produced by the dissolution of spent natural uranium and uranium oxide fuels. The HALW contains the fission products and actinides produced by the irradiation of nuclear fuels in a thermal reactor. Reprocessing extracts most of the reusable uranium and plutonium from these unwanted products. Actinides are a group of radionuclides created from the uranium by successive neutron captures (Flowers, 1976). They include thorium, protactinum, neptunium, plutonium, americium and curium. These elements have much longer half-lives than the fission products. For example, plutonium 239 has a half-life of 24,400 years. In Europe research is under way to find a way of separating the actinides from the fission products. If this is accomplished the former will be subjected to neutron bombardment in order to transform them into short-lived elements. Consequently HALW may be radioactive for hundreds rather than thousands of years. Currently the fission products and actinides cannot be liberated from each other on an industrial basis (EEC Background Reports, 1978) and it is unlikely that this will be achieved in the very near future.

Long-term storage and disposal methods are still under investigation, and may involve storage under water in ponds, in air-cooled concrete vaults, in underground geological structures and in individual packages contained in flasks (EEC Background Reports, 1978). These methods allow for retrieval of the material. HALW must be controlled safely, perhaps for thousands of years, to monitor the rate of leaching and to avoid deliberate retrieval by unlawful persons.

In advance of decisions about ultimate disposal BNFL intend to solidify (glassify/vitrify) the HALW so that it may be more conveniently managed. Without solidification HALW is mobile and containment is subject to problems of leakage. After a considerable period of storage as a solid, Britain contemplates final disposal of HALW in suitable geological formations or to the deep ocean or under the ocean bed. (For an extensive discussion, see Roberts, 1979.)

High-active waste solidification projects are under way in Europe and America and an industrial-scale glassification plant has already been built at Marcoule in France. The British programme is called HARVEST (Highly Active Residues Vitrification Engineering Studies). This process involves feeding the HALW into a stainless steel crucible and heating it together with borax and silica to form a glass. Once this operation is complete the crucible will be sealed to form a permanent receptacle.

Prior to glassification the HALW must experience a period of cooling or its intense decay heat could melt the glass. As Flowers (1976) points out, the difficulties associated with interim storage of HALW will not therefore be entirely removed by vitrification.

BNFL hope to begin vitrification on an industrial scale by the mid-1980s. The waste management procedure will then be as follows: storage of the HALW for a few years to allow it to cool, vitrification and packaging of this material for a further period of storage followed by final disposal into a stable geological formation or the deep ocean or under the sea bed (Clelland, 1978).

NOTES

1. There is no 'market' in Pu but it is possible to think of Pu as having a fuel equivalent value and hence a 'price'. The market for U is not a free one so that it is unclear if one can talk of a 'world price' for it. None the less, it is internationally traded, usually through bilateral agreements.
2. Whether Japan is able to export spent fuel for reprocessing will depend on the future attitude of the USA. Currently the USA determines that any country which is supplied with enriched uranium by her must not export fuel fabricated from the same for reprocessing without her express permission.
3. This diagram, taken from Rose, D. J., and Lester, R. K., 'Nuclear Power, Nuclear Weapons and International Stability', *Scientific American*, April 1978, has been very slightly modified.
4. Apart from LWRs other thermal reactors requiring natural or enriched uranium oxide fuel include AGRs, CANDUs, SGHWRs and HTGRs.
5. The Department of the Environment's Dangerous Goods Branch controls the transport of radioactive materials within Great Britain by road or rail. The Department of Trade are responsible for shipment of the same by sea and air. The International Atomic Energy Agency lay down the Standards that containers of radioactive material should adhere to.
6. From Colton, J. P. (1975). *IAEA Bulletin*, February 1978. 'Sealed Cask: a near-surface vault containing one or more assemblies that are sealed from air using an individual shield. Caisson: one or more assemblies sealed from air and buried near surface in the ground to use ground as shielding. Air-cooled Vaults: a collection of several assemblies within a large shielded area (building) and cooled directly by air'. Air-cooled vaults and sealed casks have been used in France and Canada respectively.
7. This diagram is a modified version of Figure I in J. P. Colton's article on 'Spent Fuel Storage Alternatives', *IAEA Bulletin*, February 1978.

REFERENCES

Allday, C. (1977)   Evidence submitted to the Windscale
                           Public Inquiry, Day 3.
Price, B. T. (1977)   Evidence submitted to the Windscale
                           Public Inquiry, Day 52.
Warner, B. F. (1977)   Evidence submitted to the Windscale
                           Public Inquiry, Day 12.
Patterson, W. C. (1976)   *Nuclear Power* (Pelican, London).
Franklin, N. L. (1975)   *Irradiated Fuel Cycle* (BNFL).
Metz, W. D. (1977)   'Reprocessing: How necessary is it for the
                           near term?', *Science*, April.
Flowers, Sir Brian (1976) Royal Commission on Environmental
                           Pollution, Sixth Report, *Nuclear Power
                           and the Environment*, Cmnd 6618
                           (HMSO).
Conroy, C. (1978)   *What Choice Windscale?* (FOE/CONSOC,
                           London).
Colton, J. P. (1978)   'Spent Fuel Storage Alternatives', *IAEA
                           Bulletin*, February.
Rose, D. J., and Lester,   'Nuclear Power, Nuclear Weapons and
    R. K. (1978)   International Stability', *Scientific
                           American*, April.
Surrey, A. J. (1973)   'The Future Growth of Nuclear Power' –
                           Part I, 'Demand and Supply', *Energy
                           Policy*, September.
Fox, Mr Justice (1977)   Ranger Uranium Environmental Inquiry,
                           First Report (Australian Government
                           Publishing Service, Canberra).
Tolstoy, Professor I.   Letter to *The Times*, 20 May 1978.
Times, The (1978)   'Committee to advise Minister on Nuclear
                           Waste', House of Commons Report, 17
                           May.
Clelland, D. W. (1978)   'The Management of Radioactive Wastes
                           from Reprocessing Operations', a paper
                           given at the Conference on the Effective
                           Management of the Nuclear Power Fuel
                           Cycle – European Aspects of the Tech-
                           nology and Economics of Uranium
                           (London, 22 March).

Clelland, D. W. (1977)   Evidence submitted to the Windscale Public Inquiry, Days 16 and 17.

EEC Background Report (1978)   'Making Nuclear Energy Safer'. 'The Community's Nuclear Safety Research Programme' (London).

Hazelhurst, P. (1978)   'U.K. signs £500m nuclear deal with Japanese but US approval still needed' *Times*, 25 May.

Roberts, L. (1979)   'Radioactive Waste – Policy and Perspective', *Atom*, vol 267, January.

7 The Windscale Inquiry: the Proceedings

INTRODUCTION

Windscale is a site on the coast of Cumbria in the north-west of England (see Figure 7.1). It contains four 50-MW reactors (the 'Calder Hall' reactors) and an early 33-MW prototype AGR reactor. Far more significantly, it is the site at which fuel from the magnox reactors in the UK is reprocessed and where plutonium fuel for military use and for use in the experimental fast reactor at Dounreay in Scotland is fabricated.

Windscale has been the site of a waste nuclear fuels reprocessing plant since 1952. From 1952 to 1964 the material was dealt within Building 204 (plant B204). In 1964 plant B205 succeeded B204. These plants had as their main function the receipt of spent fuel from nuclear reactors (military at first, then civil) and its 'treatment'. Treatment involves cooling in storage ponds until the next stage is considered. When that stage is depends on the type of fuel under consideration. Because magnox fuel cannot be stored for lengthy periods it is stripped of its cladding in the reprocessing plant and separated into (i) re-usable uranium; (ii) re-usable plutonium; (iii) low-radioactive waste which is discharged to the environment or buried on land; (iv) medium-radioactive waste which is stored ready for disposal at sea; and (v) high-radioactive waste which is stored in special containers (high-active waste tanks or 'HAWs'), as discussed in the previous chapter.

The subject of the 1977 Windscale Local Public Inquiry (WPI) was the request by BNFL for facilities to reprocess *oxide* fuels in a thermal oxide reprocessing plant (THORP). These fuels are from the later generation of reactors (AGRs) and consist of enriched uranium oxide clad in stainless steel. The plant at Windscale was adapted to deal with LWR oxide fuels in 1969, but closed in 1973 after an accident. The dispute at WPI centred on (a) the volume of fuels that would have to be dealt with, and (b) whether such fuel, whatever the volume or type, should be reprocessed or stored, or reprocessed in such a way that plutonium is not separated. Issue (a) clearly depends on the commit-

FIG. 7.1 Britain's nuclear power stations operating or under construction, 1978
(The SSEB and CEGB have been given permission to construct an AGR station
each at Torness and Heysham respectively. Construction has not yet started.)

ment of the UK to a programme of nuclear power and the mix of
thermal and fast reactors. It also depends on the extent to which the
Windscale plant processes foreign spent fuel from reactors in
Scandinavia, Europe and Japan. Issue (*b*) arises because the cladding
for oxide fuels is such that longer storage is possible. If it can be stored,

then it does not have to be separated; and if it is not separated then the amount of plutonium 'available' in the economy is reduced in the sense that it exists only 'in' the spent fuel and cannot therefore be used. What was disputed were the technicalities and economics of storage versus reprocessing, radiation hazards and the risks of proliferation. If storage was opted for then the plutonium would not be available to terrorists or as a fuel for fast reactors or, for that matter, for thermal reactors. Again, therefore, the dispute centred on whether the UK 'needed' plutonium as a fuel so badly that this need would override the problems of handling, transporting and safeguarding plutonium. That these problems exist *now* is not generally disputed. Much of the debate was concerned with an assessment of the technical change that would occur in the next decade or so for overcoming these problems.

Both the issue of the scale of the proposed plant and the issue of storage therefore related to the demands for fuel by the domestic nuclear programme. They also clearly related to other matters. Initially, however, we may observe only that, in terms of the 'nuclear debate' and 'how to decide' on an energy future or a single energy investment, energy policy was directly relevant to the WPI. It was discussed. Whether it was linked clearly to the proposal for THORP we discuss later in this chapter.

It is no part of the aim of this chapter or the next to debate whether the Inquiry Inspector, Mr Justice Parker, and his two advisers (assessors), Sir Frederick Warner and Sir Edward Pochin, reached the 'right' conclusion in recommending, firmly, that THORP be allowed to proceed without delay. (That recommendation is contained in the Inquiry Report published in 1978 – Parker, 1978.) Our concern is with a different issue, namely, whether the inquiry process was 'efficient' in terms of the criteria laid down broadly in Chapter 4. It is essential to recognise that in making that assessment, which we do in the following chapter, *the inquiry is not divorced from the events which preceded it or which followed it.* There is, for example, as much interest in the analysis of the notice given to the participating parties for the inquiry as in the actual proceedings, as much interest in the way in which the inquiry's findings were conveyed to the relevant Minister (i.e. the Inquiry Report) as in what happened in the 100 days of the inquiry, and so on. That the whole process is seen as a 'system' is important because it is the efficiency of the system that is under scrutiny. It is also important because, as noted in the preface, the nuclear debate is characterised by some remarkable attitudes on both sides. Criticism of a process is not therefore a criticism of the individuals who find themselves part of that process. Clearly, a disagreement can exist if they defend that process as Mr Justice Parker has done publicly.[1]

THE TERMS OF REFERENCE

In a written answer in the House of Commons on 18 May 1977, Mr Peter Shore, Secretary of State for the Environment, said of the Windscale Public Inquiry, 'Planning inquiries do not have "terms of reference" as such, but, in accordance with the inquiries procedure rules, I issued a statement on April 5 setting out the points which, on the information so far available, appear to me likely to be relevant to my consideration of the application.'

This said:

The Secretary of State will consider all relevant aspects of the proposed development. On the information so far available the following points appear to him likely to be relevant to his consideration of the application:

1. The implications of the proposed development for the safety of the public and for other aspects of the national interest;
2. The implications for the environment of the construction and operation of the proposed development in view of the measures that can be adopted under

   (i) the Radioactive Substances Act 1960 to control the disposal of solid, gaseous and liquid wastes which would result from the proposed development; and
   (ii) the Nuclear Installations Act 1965 to provide for the safety of operations at the reprocessing plant;

3. The effect of the proposed development on the amenities of the area;
4. The effect of additional traffic movements both by road and rail which would result from the proposed development;
5. The implications of the proposed development for local employment;
6. The extent of the additional provision that would need to be made for housing and public services as a result of the proposed development.

This is the 'Rule 6' statement from the Secretary of State and it was this that was designed to guide Mr Justice Parker in his deliberations.

THE TIMETABLE OF EVENTS

Table 7.1 sets out the timetable of events prior to and succeeding the WPI. Some other succeeding events are referred to in the text. Several comments may be made about the timetable of events.

First, one commentator, writing after the publication of the Parker Report (Conroy, 1978), points out that but for the activities of certain pressure groups 'the Inquiry very nearly did not take place' (p. 6). Exactly what happened in 1976 is not clear but it seems fair to say that there was considerable pressure on the Secretary of State to 'call in' the application. Table 7.1 shows that as late as November 1976 the Secretary of State for the Environment (SOSE) had still not decided whether to call in the application, even though, technically, the full 21 days in which he should make that decision after it had been referred to him by Cumbria County Council had lapsed. Various statements and theories about what happened exist but the silo seepage at Windscale in the previous October – which was not made public until December – clearly led to various further pressures. There was extensive pressure, from without and within Parliament. The climate of opinion in late 1976 was therefore such that the Inquiry could very easily not have happened at all.[2] This reflects on the 'monitoring' of local planning applications by the Department of the Environment. In future, it would seem sensible for planning applications which have more than a local dimension to be monitored to see if they are best dealt with by means other than the local planning inquiry. It would be a fairly simple matter to advise local authorities to pay special attention to applications to see if they merit the *possibility* of being placed on the 'national agenda', using a set of guidelines to this effect.

TABLE 7.1   Timetable of events

1952–64	B204 plant operates at Windscale (magnox).
1964 to date	B205 plant succeeds B204 (magnox).
1969	B204 converted to deal with oxide fuels.
1971	BNFL assumes responsibility for Windscale.
1973	Closure of oxide reprocessing plant due to accident.
1974	BNFL announces plan for THORP.
1975	BNFL negotiates with Japan for reprocessing Japanese spent fuel.
1976 (January)	'Church House' Debate between opposition groups and BNFL. Secretary of State for Energy attends.

1976 (12 March)	Government announces support for BNFL Japanese contract negotiations.
1976 (April)	Friends of the Earth hold open-air debate at Windscale.
1976 (May)	Cumbria County Council announce that any BNFL application to expand plant at Windscale would require prior public consultation.
1976 (June)	BNFL submit planning application to Copeland Borough Council. Referred to Cumbria CC. Application is for new oxide fuel reprocessing plant, plus modifications to magnox reprocessing plant plus vitrification research plant.
1976 (September)	RCEP 6th Report published. Whitehaven public meeting organised by Cumbria CC.
1976 (2 November)	Cumbria CC give approval to BNFL application but draws attention of Secretary of State for Environment (SOSE) to it as possible departure from approved County Development Plan.
1976 (24 November)	SOSE calls for more time to decide whether to 'call in' BNFL's application. Seepage of radioactive material from a silo at Windscale made public.
1976 (22 December)	SOSE announces that BNFL should resubmit their application in three parts and that there would be a public inquiry into that application dealing with oxide fuel reprocessing.
1977 (21 January)	BNFL amend Planning Application to exclude oxide reprocessing plant.
1977 (24 February)	BNFL prepare separate Planning Application for oxide reprocessing plant.
1977 (March)	BNFL submit THORP proposal to Copeland Borough Council who refer it to Cumbria CC. Application 'called in' by SOSE, under section 35 of the Town and Country Planning Act of 1971, on 25 March.
1977 (31 March)	Mr Justice Parker appointed Inspector and Professor Sir Frederick Warner and Sir Edward Pochin as assessors.
1977 (1 April)	Official Notification of location, date and time of WPI.
1977 (5 April)	SOSE states terms of reference for WPI.
1977 (6 April)	Notification of 17 May Pre-Inquiry Meeting.
1977 (2 May)	SOSE informs BNFL about objectors and their objections.
1977 (11 May)	BNFL issues summary of case it will present.
1977 (17 May)	Preliminary meeting of inspector, BNFL and objectors to discuss procedural matters.
1977 (14 June)	Opening of WPI
1977 (4 November)	Closure of WPI.

(Table 7.1 *contd.*)

1978 (26 January)	Mr Justice Parker presents his report to SOSE.
1978 (6 March)	Parker Report published. SOSE announces formal 'refusal' of planning permission for BNFL.
1978 (22 March)	House of Commons debates Parker Report.
1978 (3 April)	The Town and Country Planning (Windscale and Calder Works) Special Development Order 1978 is laid before the House of Commons.
1978 (15 May)	Windscale and Calder Works Special Development Order 1978 debated in the House of Commons. Majority of 244 to 80 in favour. Establishment of Radioactive Waste Management Advisory Committee under chairmanship of Sir Denys Wilkinson.
1978 (24 May)	Sir John Hill, Chairman of BNFL, signed the £500m contract for transportation and reprocessing of Japanese spent fuel.

Some objectors at WPI complained of the very short space of time between the calling in of the application (25 March 1977) and the opening of the WPI (14 June 1977), a space of under three months. Indeed, most parties, including BNFL, would have appreciated greater notice. Without wishing to favour one group or another, it is the case that many objectors lack funds and full-time personnel to represent them at an inquiry. If they have to attend themselves they may have to arrange time off work, adjust holidays and so on. Three months is not adequate notice to make even such simple administrative arrangements and many objectors suffered accordingly. There is of course a contrary argument, that if an objector is so moved by an issue he or she will manage to attend somehow and to make his or her views heard. It is unlikely, however, that this is the proper way to secure a debate, or even a representative exchange of views. In future, a significantly longer period should be given for all parties to prepare their case. Postponements of a year would seem to be too long if employment issues or contracts are involved, though small groups of objectors could probably benefit from even longer periods of notice. Six months between the publication of the terms of reference (and the applicant's case for the proposal) and the start of the inquiry would seem to be a minimum. Mr Justice Parker did not agree that there was an inadequate period of preparation. He points out that the SOSE's decision to call an inquiry was announced in Parliament on 22 December 1976 and 'there were thus six months in which to prepare. I consider this period to be amply sufficient' (Parker, 1978, p. 80).

Reference to Hansard for November 3 1976, the day after Cumbria CC decided to refer the BNFL application to the SOSE, reveals an exchange between Mr Shore (SOSE) and several MPs. It seems significant that one of Mr David Steel's questions – why did the SOSE 'not call in this application in the first place since the implications of the decision go far beyond the interests of Cumbria alone?' – was not answered in the exchange. Mr Shore did refer to 'a public meeting in Cumbria' (the September 1976 meeting at Whitehaven) at which 'the overwhelming opinion . . . was in favour of going ahead with the project . . .'. Reference to the national dimension by Mr Robin Cook also elicited no response. It seems fair to say that, at this time, the Secretary of State had not considered the national implications of Windscale in any detail or had other reasons for not calling in the application.

But Mr Justice Parker's assessment of the situation in respect of timing is not quite adequate. What the SOSE announced on 22 December 1977 in Parliament was the fact that he was referring the matter back to BNFL so that they should submit their previously tripartite proposal in three separate aspects so that, in turn, he could then call in that proposal relating to oxide fuel reprocessing. He did indeed promise an inquiry into that proposal and made it clear that it would be a *local* public inquiry and *not* a planning inquiry commission. Had the latter been selected, it would probably have triggered off an understanding on the part of many that *national* issues would be debated. Yet *the terms of reference for the WPI were not stated on 22 December* nor did the Parliamentary exchange indicate in any way that national aspects would be debated. The very choice of a *local* inquiry reinforced this negative aspect of the exchange. It is technically correct to say that the time-scale between the announcement of the inquiry and its opening should be dated from 22 December. Anyone monitoring their Hansard would certainly have known of the decision, but not what the inquiry would debate. They certainly would not have known the terms of reference, since they had not been decided at that point *and were not known until 5 April 1977*, only two months before WPI began.[3]

A check of the major newspapers for 22 and 23 December shows that on 22 December, the day of the SOSE's announcement, Anthony Tucker of the *Guardian* reported the likelihood of an inquiry and a reaction from Friends of the Earth. On 23 December Tucker ran a special piece in the *Guardian* on the SOSE's announcement of 22 December. That piece, entitled 'Windscale is put on trial', gives no indication of the scope of the inquiry, nor could it have done. It repeats parts of the SOSE's statement and a reaction from BNFL, including an odd statement to the effect that 'it now seems possible we may never go ahead with the oxide plant proposals'. Both Tucker's piece and the report by Pearce Wright in the *Times* pointed out that there would only by an inquiry *if* BNFL submitted a separate application for the oxide reprocessing plant. This was, in both cases, correct reporting of what the SOSE said on December 22. The *Times* report also gave no indication of the scope of any inquiry nor did it speculate as to what it might be.

It seems clear that, while the strong *probability* of an inquiry was known about on 23 December 1976, considerable uncertainty remained since (*a*) the SOSE had said there would be an inquiry only if he received a separate application; (*b*) someone at BNFL reported to Anthony Tucker that the reprocessing plant might not go ahead at all; but, most important, (*c*) the scope of that inquiry, if it took place, was not known. As noted above, even BNFL did not know what case it was to prepare until well into 1977. The argument that six months' notice was given *de facto* therefore seems misleading.

A further observation concerns the meeting on 17 May in Whitehaven at which the Inspector met representatives of BNFL and some of the objectors. These pre-inquiry meetings are fairly standard and are used to 'organise' the proceedings. It is an oddity that the records of that meeting show that Mr Justice Parker was at that time – less than a month away from the inquiry – saying that he was not clear how far 'the question of the need for the proposed development would fall within the scope of the inquiry'. Again, this is a legitimate confusion given the vagueness of the SOSE's first term of reference, which spoke of 'the safety of the public' and 'other aspects of the national interest'. The first half of this statement should certainly have raised the issue of the transportation of waste fuels and separated fuels between nuclear sites. There is no alternative to including this aspect of the debate given that statement. The second part of the statement is vague in the extreme, however, and it would be understandable if no party construed this to mean that there was a question of whether reprocessing should take place *at all*. Accordingly, it seems that as late as 17 May the questions to be discussed at WPI were still unclear to all parties, including the Inspector. This is not a satisfactory state of affairs for a public inquiry debate and it points to the need to issue clearer statements for 'terms of reference' where national issues are involved.

There are arguments against issuing such clear terms of reference. Technically, inquiries do not have terms of reference and an explicit flexibility is built in so that the Inspector may determine what is relevant as the inquiry proceeds. However, such flexibility may still be retained in the context of clearer statements of what is required by way of a report on national aspects. Further, vague terms of reference aid neither the Inspector, the appellant party nor the objectors. In short, they make a fruitful debate all the less likely.

Some time between 17 May and 14 June, the opening of the WPI, the Inspector had in any event decided that 'need' would be debated (thus departing from all precedent in local public inquiries), since his very first question was whether oxide fuels should be reprocessed in the United Kingdom at all.

It would also seem to be the case that pre-inquiry meetings should serve a more important function than they do. Exactly how they are used appears to be a matter for the Inspector and the parties present. They establish procedure, hours for the hearings, and so on. What might well have been done in the WPI case was to have secured a far better organisation of evidence so that overlapping and repetitive evidence was avoided. It would also permit inspection of each party's main evidence. This issue is taken up in more detail in Chapter 10. The WPI pre-inquiry meeting seems to have been held far too late (a result of purely timetabling factors once one looks at the timetable in Table 7.1) and in the context of terms of reference which still needed clarification. On what basis evidence should be presented, e.g. in sequence by party or by subject, is discussed later.

The use of the pre-inquiry meeting to set, as far as possible, a timetable for the presentation of evidence also bears on views about the 'compression' of the Inquiry into 100 days. It was felt by many objectors that there was not sufficient time *within* the Inquiry timetable to digest new material, and, moreover, that the hurried sequence of events also prevented greater public awareness of the issues. In the latter respect, it is not clear if, say, several breaks of one week would have permitted the public to learn more of the proceedings or not or given participants time to 'digest' material. To some extent, what take-up of the proceedings there was in the media could be held to have been sustained by the fact that the proceedings were a 'non-stop marathon'. That the speed of the proceedings was such that parties to the Inquiry did not have time to prepare adequately for cross-examination, etc., seems without question. This complaint has come from both BNFL and objectors. Again, therefore, it is not a criticism emanating solely from the 'losing side'. It seems that, had the notice for the Inquiry been adequate and had the parties known which questions were relevant to the debate this 'compression' effect could have been reduced considerably. Ironically, it might even have reduced the actual length of proceedings and their cost.[4]

THE PARTICIPANTS

Appendix 1 presents a comprehensive list of the personnel who took part at the WPI. As far as possible, these have been classified with their affiliations in terms of whether they were 'pro-THORP' or 'anti-THORP'. We have also tried to summarise what each person's view was. Appendix 1 is presented largely for informational purposes and as a kind of 'index' of the WPI. It is there largely for its own interest. We would draw attention to one observation, however. By far the greater number of the persons appearing at the Inquiry were 'experts' in one field or another. This is inevitable in terms of debating the technical issues at Windscale – safe radiation exposure levels, and so on. Virtually all witnesses had some professional qualification, univeristy degree and so on. Only some local objectors, and some 'impromptu' ones, seem to fall outside this classification. Given the nature of the discussion and the problem, this observation can give little cause for concern. What it does underline is the earlier statement that the 'public' is an entity larger than the people represented at WPI. It can be argued that the public, if they are interested, should have ready access to these deliberations or should themselves participate. The term 'Public Inquiry' seems to be a misnomer for what are essentially professional debates. This may be inevitable, as noted above. It may also be acceptable if ways can be found to relay the debate to a wider public in such a way that they can reasonably be held to be in a position to 'participate' in that sense.

In the 22 March 1978 debate in the House of Commons exactly this sentiment was expressed by one MP who stated that the WPI was

> a classic example of the general imperfections of the present system we laughingly call public inquiry, public accountability and grass roots democracy. These public inquiries are meaningless nonsense to people outside. The objectors are either wealthy and able to afford counsel, or middle-class and articulate. It is very difficult for us to gauge the working-class reaction. (Hansard, 22 March 1978, col. 1630).

The difficulties that confront any attempt to broaden participation are readily acknowledged. Nevertheless, the attempt is worth making.

THE ARGUMENTS

What follows is an overview of the salient arguments put at the WPI. No attempt is made to suggest which arguments are correct and which are not, since our concern is with the nature of the debate and the way in which information was 'channelled' through the Inquiry to the public. The 'issues' are presented in no particular order so that no implications should be imputed to the sequence.

(a) Routine Releases of Radiation

No nuclear fuel cycle can operate without *some* release of radioactive substances. The ones of interest are those that are, as it were, a necessary consequence of the technology – the so-called 'routine' discharges from power plants; those that arise from accidents; and those that may arise from the eventual disposal of waste from the reprocessing facility. As far as routine emissions are concerned, two major issues arise. First, what will the level of emission be and will the operator, in this case BNFL, be able to honour any agreed standard? Second, whatever the standard what is the damage associated with it given that *any* radiation is held to be harmful?

The level of expertise brought to bear on these two questions was high, particularly on the latter. Two types of harm are distinguished for radiation: somatic harm, the damage done to an individual, and genetic harm, the harm done to descendants through exposure to radiation by predecessors. The units for measuring radiation levels vary but in the UK wide use is made of the 'rem' – the 'Roentgen Equivalent, Man'. This unit depends on the effective radiation absorbed by body tissue and is in turn a function of two variables, the amount of energy (the 'dose') delivered and the quality or nature of the radiation. The former is often measured in 'rads' ('Radiation Absorbed Dose').

The 'rad' is equal to the amount of radiation that will cause 1 kg of material to absorb 0.01 joules of energy (radiation imposes an 'energy impact' on materials and tissue). But one rad of β or γ radiation is less biologically harmful than 1 rad of neutron radiation or α radiation. The ratio of α and neutron radiation to β and γ radiation use is 10:1. So, a weighted product of the rads (the 'dose') and the 'quality' (α, β etc.) gives us the 'rem'. Therefore 5 rems could be made up of 5 rads of γ radiation, or $5/10 = 0.5$ rads of α radiation. Clearly, assigning health significance to the radiation from the routine operation of THORP is meaningless to other than those expert in the public health field. Mr. Justice Parker revealed a concern to explain these health risks in terms that could be 'pictured' by the general public. Thus, exposure of a worker in THORP to 5 rems per year is, on the basis of UK interpretation of the evidence, thought to be associated with a one-in-2000 maximum risk of a fatal cancer at some point in the exposed person's life. This is roughly the same risk as dying of an accident in a single year (Parker, 1978, para. 10.32). If people were exposed to 1 rem per year, BNFL's intended maximum for workers at THORP, the risk would be one in 10,000, or the same risk of contracting cancer by smoking three cigarettes a week. (See Appendix to this chapter.)

The 5 rem and 1 rem illustrations are important. The latter represents what BNFL have committed themselves to observing as a maximum for workers, although whether they can achieve it or not is another matter. The former is the maximum worker dose suggested by the International Commission on Radiological Protection, an international body of experts that, technically, has no statutory authority but which, as several witnesses pointed out at WPI, had effectively become the standard-setting authority here, although this latter task belongs strictly to the National Radiological Protection Board. Since NRPB and ICRP are not wholly independent of each other, however, it is natural that the standards set by the international body should influence those set by NRPB.

In the same way, exposure of the population at large will lead to cancers. BNFL's stated *intent* was that public exposure would not exceed 50 millirems, where one millirem is 1000th of a rem. In short, the aim would be to keep public exposure to one-twentieth of that for workers. The ICRP limit for public exposure is 500 mrems, so that, if the intent is honoured, THORP would expose the public to one-tenth of the ICRP maximum, at most. These standards are supposedly set so as to be *exclusive* of medical exposure through X-rays. They also relate to what is termed 'whole-body' exposure. Quite different levels are set for specific organs of the body, for the thyroid gland and even for the hands and the feet. Further, the limits set are for design purposes and actual doses will be lower.

The 'quality' of radiation was seen to be one of the determinants of the ICRP standards as expressed in rems. One example can be cited since Mr Justice Parker went to great pains to counter what he considered was the 'emotionalism' attached to plutonium which, as Chapter 6 explained, would be isolated as a fuel in THORP and which would be the dominant fuel in the event that fast reactors were introduced. Plutonium has a long half-life, it takes many years to decay. Pu 239 in fact has a half-life of 24,400 years. If ingested – i.e. swallowed or taken into the body other than through inhalation – most of it will be released through the body. Its long half-life means that its radiation level is low. None the less its biological effect is considerable. Being an alpha particle emitter, Pu 239 is regarded as having some ten times the biological impact of beta or gamma particle radiation. If inhaled, as opposed to being ingested, Pu exits from the body more slowly and can be retained long enough to cause tissue damage to the lungs and, thereby, increase the risk of lung cancer. It matters therefore not just what the source of radiation is, but the route by which it may reach the body or its organs, and where in the body it 'resides' until finally lost.

Now, although it is valuable to express risks from radiation in 'comparable picture' terms such as the everyday risk of an accident, it does run an important risk of misleading people, namely, in suggesting to them that these equal risks are equally acceptable. It is widely felt that an involuntary risk – i.e. one 'imposed' on the person – is regarded as being much less acceptable than a voluntary risk – one that is freely entered into. Starr (1976) concludes that 'social demand' calls for safety levels 1000 times higher for involuntary risk than for voluntary risk. A radiation worker would therefore be able to treat the two risks as directly comparable if it could be argued that he or she 'freely' worked at a reactor or reprocessing plant. That public might wish to multiply the risk to themselves by some factor to represent the fact that they have not voluntarily accepted the risk except in the indirect sense of having acquiesced in the investment decision. This point was put at WPI. Mr Justice Parker's assessment was that the public would be willing to accept the benefits of THORP for the risks associated with a 50-mrem maximum exposure p.a. That risk is said to be 1 in 200,000 of a fatal cancer in Mr Justice Parker's analysis. This excludes risks of other damage but it was widely thought that genetic damage would not be entailed at this level, although, again, witnesses disagreed.

Relating such a finding in 'picture' terms to what the public might think the expression of risks therefore has advantages *and* disadvantages. It remains unclear (a) how the public's own assessment of the acceptance of involuntary risk could be made known to the Inquiry, and (b) how the public was to convey its thoughts on the acceptability of even a voluntary risk of fatal cancer equal to 1 in 200,000. There can be no suggestion that it was part of the Inquiry's task to assess public opinion in this way. None the less, it remains the case that a surrogate opinion was made by suggesting that the public would accept such risks (Parker, 1978, para. 10.34). The Inspector had no option but to make such a judgement, but it is quite arguable that far more exposure to documentation and study would assist the public in conveying its view beforehand rather than having it voiced by a system which is expressly non-consultative.

The issue becomes infinitely more complex when it is recognised that serious scientific dispute exists over the effects of radiation, particularly low-level doses. A sequence of eminent experts appeared to argue that the ICRP limits were greatly in excess of a desirable maximum (see Appendix 1 and the views of Professor Radford, Dr Stewart, Dr Ichikawa, Dr Spearing, Professor Tolstoy). Others criticised monitoring procedures, especially for food chains such as fish, and the limits set for caesium in fish. Much of the debate centred on (i) the fact that the ICRP limits were laid down in 1959 and had not been changed in the light of recent work, (ii) the suggestion that the linear dose hypothesis was wrong, and (iii) the apparent contrast between USA practice in terms both of institutions and standards.

ICRP Publication 26 was released in August 1977, during the Inquiry. It continued to recommend the maximum standards laid down in its second publication, ICRP 2, in 1959. It was the argument of some of the witnesses, notably Dr Stewart and Professor Radford, that these standards failed to acknowledge recent work in the field. This recent work, it was argued, pointed to the fact that low doses were more damaging than had hitherto been considered. Much of what Dr Stewart had to say rested on her own study of workers at the Hanford reactor plant in the USA. This study was controversial in its history, being published in 1977 but having been initiated by Dr Thomas Mancuso in 1964. Professor Mancuso is Professor of Occupational Health at Pittsburgh University but his contract with the Atomic Energy Commission and Energy Research and Development Association (subsequently ERDA) was terminated in 1977. In the late stages of this project Dr Stewart had assisted in the work, jointly with Dr Kneale. The study has itself been the subject of fierce debate, mainly on the statistical methodology used. The interested reader is referred to the original paper (Mancuso *et al*, 1977) and to two criticisms (Reissland and Dolphin, 1978, and Sagan, 1978). If it is correct, however, it could cast some doubt on the 'linear hypothesis'. Under this hypothesis, cancer incidence is related to radiation doses on the assumption of a linear relationship which, moreover, is of a proportional form – i.e. the straight line, if it were drawn, would run through the origin. This hypothesis had always been considered 'cautious' in that it effectively extrapolated backwards to the origin from observations on fairly substantial doses of radiation in human populations exposed to radiation. In short, unlike the damage functions one might expect from the usual toxic effects of chemicals, whereby small doses of a pollutant may not generate any damage at all, the linear hypothesis effectively assumed that *some* damage will be generated. This is now an assumption generally applied to all environmental carcinogens. If this assumption overestimates the risks, standards based on it could be assumed to be 'generous' and over-restrictive on the industry.

The Hanford study was suggesting something quite different, namely that low levels of radiation, such as those experienced by workers at the reactor, were responsible for an excess number of cancers as much as 20 times more than the linear hypothesis would predict as a maximum feasible number. On this basis the linear hypothesis was not 'conservative', as the Hanford study was showing effects greater than had been expected. In the event, Mr Justice Parker clearly recognised the import of Dr Stewart's evidence, in no small part because the Hanford study was the only study on a large scale of workers in the nuclear energy industry. Evidence by Stewart and Kneale occupies some 70 pages of transcript. In the event, Mr Justice Parker supplied his own criticism based on the work of others and concluded that the difference between cancer and non-cancer deaths at Hanford had not been shown to be statistically significant.

Professor Radford made very much the same suggestion as Dr Stewart about low-level radiation but based his conclusion on a number of recent studies and his own experience with the USA BEIR (Biological Effects of Ionising Radiation) Committee, of which he is Chairman.

It is perhaps useful to rehearse some of the history of the BEIR Committee since it bears on (*a*) the status of Professor Radford and (*b*) the relationship between BEIR and ICRP. Essentially, ICRP grew up as a sister organisation of the US National Committee on Radiation Protection (NCRP), with fairly extensive common membership. Until 1969 it seems fair to say that ICRP was 'dominated' by the US, the UK and Canada. At about that time, however, the nuclear controversy became strong in the USA and the NCRP came under criticism for bias, not least because it was self-perpetuating, not technically an official organisation, and supported by the nuclear industry as well as medical radiology societies. To get out of this impasse Congress, then faced with an incipient Environmental Protection Agency, called for an independent body on radiation to be convened by the US National Academy of Sciences. By inference, the criticism of NCRP was criticism of ICRP due to overlapping membership. BEIR therefore began its work in a rather different context from ICRP and NCRP and, indeed, caused an early uproar in 1972 by criticising medical radiological practice. It also seems fair to say that NCRP is no longer much regarded in the USA.

The history seems important because, whatever the format of an inquiry, personalities intervene and what Professor Radford, and other witnesses, did was to question the very role of ICRP and the validity of its work as a basis for setting standards. Yet one of the assessors, Sir Edward Pochin, was, and is, a prominent member of ICRP.

The major suggestion made by Professor Radford was that the whole body dose limit should be 25 mrems and not the 500 mrems suggested by ICRP. His testimony took up over 90 pages of the transcript. In the event, Mr Justice Parker's report accorded Professor Radford's arguments for a 25 mrem standard one paragraph, five other paragraphs being related to his proposals and ancillary suggestions. Radford expressed concern that the latest document, ICRP 26, published during the Inquiry (ICRP, 1977), exhibited no assessment of recent literature, which, in his view, gave rise to concern that low levels of radiation were more damaging than had hitherto been thought. Interspersed with his scientific assessment of the evidence was comment on the status of ICRP and the way in which its recommendations had become the norm in many countries. ICRP is not 'responsible' to any governmental authority and its composition is based on self-selection by existing members whenever a vacancy occurs.

The 25 mrem p.a. maximum standard was not simply a personal recommendation on Radford's part. It was already embodied in US Environmental Protection Agency regulations (US EPA, 1977) to become effective as from 1979. This difference in standards between two major nuclear power countries receives no mention in the Inquiry Report and, on the basis of the requirement laid down earlier that maximum information be imparted to the public at risk, it seems odd that it went largely unnoticed or commented upon. The issue is now not just who is correct in a scientific dispute, but also why scientific opinion in favour of arguing that low-dose concentrations are more important than hitherto considered has had an impact on one standard-setting government agency but not on another. Some suggest that the National Radiological Protection Board in the UK are too ready to take ICRP assessments at face value, even though those assessments had not changed in nearly twenty years.

The other difference would seem to arise from the way in which standards are set. As noted above, any radiation has some effect, so that where one 'draws the line' depends on the assessment of cost (in terms of health damage), the money cost of achieving given standards and the assessment of benefit from the relevant nuclear installation. In short, someone has to carry out a 'risk benefit' analysis. In the UK this analysis has been subsumed under the requirements that ICRP maximum limits should not be exceeded, *irrespective of cost*; that the whole population should not receive an average 'genetically significant' dose in excess of 1 rem per person in 30 years from radioactive waste disposal; and that everything that is 'reasonably practicable' should be done to keep levels below these maxima. In practice this has meant that aqueous discharges have had numerical limits attached to them while ambient discharges have not, the latter being subject to agreement on what is reasonably achievable. Just how an engineer is to interpret such requirements is unclear. However, the EPA regulations covering reprocessing plant cover *ambient* releases only – i.e. they assume *no* release of radioactive content to the sea or other waters. For this reason it is held that the two standards are not comparable. 'The existence of both liquid and airborne effluents [in the UK] invalidates a direct comparison of the UK authorisations with these EPA standards' (Richings *et al*, 1978). Further, the EPA regulations are specific to nuclear sources whereas the ICRP standards cover all sources of radiation. Lastly, the EPA standards cover the light water reactor cycle with some exceptions such as waste disposal plant.

Are the two standards therefore the same or nearly the same? According to one report (NRPB 1978) the ambient emission standards which BNFL aims to achieve at Windscale are 'within or slightly exceed the EPA limits' (Richings *et al*., 1978). The krypton 85 releases would exceed the EPA limits by a factor of six, but since one of the conditions laid down for constructing THORP has been the introduction of a krypton removal plant, this comparison ceases to have validity *if* the plant is successfully introduced. Note that *existing* emissions (33.5 mrem) exceed the EPA limits. Whether the EPA limits and the ICRP limits are similar seems to be a matter of interpretation. As presented by Professor Radford at WPI they are distinctly different and the EPA limits exceed the ICRP limits by a factor of 20 for whole-body doses. Further, bearing in mind that standards must necessarily be set with reference to cost-effective approaches, it is not clear what to make of the NRPB 1978 statement that 'Evaluations are sensitive to the various value judgements, particularly on the weight to be attached to very small doses to large numbers of people and the international impli- cations of this'. That the same philosophy, cost-effectiveness, is being applied is not in dispute. That it should apparently lead to numerically different standards in different countries would seem, in part anyway, to reflect the differing value judgements about what death rate is 'reasonable' given the benefits of nuclear power. But if value judgements differ it would be useful, indeed necessary, to see them spelt out to the people who actually bear the risk and who may not collect the benefit.

In April of 1978 the NRPB recommended standards for levels of radiation from waste discharges in the nuclear industry. These standards were set at 500 mrems for critical groups (those near installations or with particular eating habits leading them to consume, say, contaminated fish), which would mean an average of 100 mrem per year or some 7 rem for a whole-body dose equivalent over a lifetime. For the general population 50 mrem is thought an appropriate annual average for *all* sources of radiation. It is thought that doses from nuclear waste disposal would account for only 5 mrem of this annual average. If BNFL succeed in keeping radiation levels from THORP down to 50 mrem or below they will therefore be well below the recommended *critical group* NRPB standard. In June 1978 the NRPB announced that it would be compiling a register of deaths of persons employed in AEA, BNFL, CEGB and Ministry of Defence Establishments and would maintain records of their exposure to radiation. It is expected that the records will cover 90 per cent of those employed after January 1976 and in employment after that date. Consideration is also being given to an analysis of 100,000 past workers. The aim is to establish whether low-level radiation is or is not important. Such an announcement at least suggests that the difference in scientific opinion on the effects of low-level radiation is a real dispute. It was noted earlier that the evidence used to set standards was almost entirely based on information unrelated to the actual exposure of nuclear industry workers. Of course, unless the retrospective studies succeed, it will be upwards of 20 years before the debate will be resolved, if it is, by such a study.

It was perhaps an irony that not long after this study was announced, the Ministry of Defence announced that 'excessive' amounts of plutonium had been found in the lungs of workers at their weapons establishment in Aldermaston. The first discoveries had in fact been made in January of 1978 but became public in August. The Government acted swiftly by appointing Sir Edward Pochin, one of the assessors at Windscale, to investigate the findings. His report was submitted to the Minister for Defence in November 1978 and it concluded that, while contamination was less than originally feared, the Aldermaston safety standards did need tightening. Opponents of nuclear power were quick to seize on these findings as evidence that, whatever the promise made by a nuclear establishment it may not be able to keep it. One essential difference however is that the Aldermaston site had not engaged in whole-body monitoring whereas the civil nuclear power industry had.

The general conclusion must be that, given the serious dispute over low-level radiation effects, the public have a right, as have industry workers, to expect caution in the setting of standards, and especially to know on what assessment the standards are based. In his Report Justice Parker felt that the ICRP standards were sound and that, in any event, BNFL would honour their promise to keep public exposure below 50 mrem even if they were not meeting that target now. Some other aspects remain and require only brief comment. In dealing with the evidence of Professor R. Ellis on radiation, Mr Justice Parker remarks that some workers had in the past received radiation doses 'at or near to – and in some cases above – permitted limits' and that the number who had experienced these doses 'was higher than it should be' (para. 10.61). He none the less concluded that, despite its failure in the past, BNFL would succeed in the future. Professor Radford in his evidence had stated that, in his experience, the best guide to future capabilities was the success in honouring past standards. It is arguable, but hardly convincing, that the very fact that the WPI took place at all and that the Report itself generated so much anger among objecting parties could be held to ensure that BNFL would be more than gravely embarrassed if they did not honour their promise.

The Inquiry Report also effectively endorses the ICRP standards and rejects the evidence of Dr Stewart, Professor Radford and others on low-level radiation. Someone who read only the Report and not the transcripts would perhaps be unaware of the seriousness of this aspect of the debate, especially as Professor Radford's lengthy submission is not dealt with in other than cursory terms. In the event, some reassurance exists in the fact that NRPB are now to maintain a register. The cry that it is 'too late' is inevitable and investigation of past workers would seem essential.

The Report also rejects the view that the existence of a scientific dispute is reason to delay an investment. The reasoning rests on the assumption that dispute will always exist and, hence, to use the existence of dispute as a reason for delaying investment would amount to never undertaking the investment. How far this line of reasoning should be followed must, however, depend on how serious the risks of being wrong are. That is, the costs of being right or wrong are not symmetric. To take a decision and then discover that it was based on radiation standards that were unreasonably damaging would be to make the wrong decision. To postpone the decision is of course also to incur costs in terms of forgone energy or GDP or whatever units the benefit is expressed in. Arguably, even if a 50-mrem upper limit was maintained and turned out to be above some strictly desirable limit, the costs of exceeding the limit would not be equal to the costs of the delay. Equally, the outcome might be the opposite. It is not just a matter of which line of reasoning to choose. It rests in part on pursuing a line of reasoning which spells out to the persons affected by the decision the costs and benefits *they* receive compared to the costs and benefits to the country as a whole, underlining one theme of the present report – that one purpose of the overall advisory process is to inform the general public of decisions taken in their name.

(b) Energy Futures

It was noted earlier that the decision to invest in THORP rested very much on one's view of the 'energy future'. More specifically, it depends on exactly what the nuclear component of an energy future is to be. It can safely be assumed to be greater than zero simply because nuclear power stations already exist which use oxide fuels and the spent fuels from these stations must be disposed of in some fashion. The fact that spent fuels exist and will increase in quantity (unless the AGR programme and any water reactor programme is abandoned) does not, however, necessarily entail that reprocessing must occur. For a limited nuclear programme the storage and disposal option remains, depending on the technical facts about storage. The final disposal of spent fuels may also entail finding sites for *less* material than is disposed of via the reprocessing route – see subsection (*c*) below on 'Reprocessing versus Storage of Spent Fuels'. On the other hand, the storage/disposal route could involve disposing of *more* plutonium, although exactly what that ratio is is open to dispute. It is far from clear, therefore, that reprocessing as an option is entailed, in any way, by the *existing* AGRs and those under construction. Although Mr Justice Parker indicated that he thought this was the preferred route for such a limited scenario, it is not a necessary one unless the storage/disposal arguments are totally in error.

Scenarios in which the nuclear component of future energy supply increased in size were presented at WPI, notably by Dr Peter Chapman of the Energy Research Group at the Open University and by Mr Gerald Leach (see Chapter 2). We saw that Leach's proposed nuclear contribution was 30 mtce in the year 2000. Chapman's scenario allowed for 25 GWe of installed capacity in 2000 or some 60 mtce. According to Parker (1978, para. 8.49), Chapman's scenario was consistent with a backlog of 6000 tonnes of spent fuel by 2000 and a per annum generation rate of 600 tonnes thereafter. Approximately, therefore, Leach's scenario would produce about 3000 tonnes backlog, and 300 tonnes for 2000 but gradually declining into the twenty-first century as, in his scenario, the nuclear contribution begins to be phased out. Both scenarios are consistent with a storage/disposal route or with a reduced size of plant for reprocessing. According to Conroy (1978) a smaller plant dealing with only domestically produced waste would be uneconomic.

Extensive attention was paid at the WPI to these and other alternative energy futures. The official forecasts were criticised for overstating demand in the future; the virtues of policies which matched end-use and source on a small scale were advocated, and so on. It was pointed out that the official forecasts had already been drastically reduced from 650 mtce for 2000 to 560 mtce, a change of 14 per cent in only one year, 1977 – this is for the 3 per cent economic growth target. Informed commentators expect further downwards revisions. None the less, as far as Windscale was concerned, it is the electricity component and, in turn, the nuclear component of this that matters. This had changed hardly at all during the revisions. (Critics setting out to cause embarrassment could, however, have drawn attention to the AEA's own forecasts for nuclear capacity contained in the RCEP Report in 1976. This implied 104 GW of nuclear capacity in 2000, of which 33 GW comprised fast reactor output!)

For information, Figure 7.2 shows electricity consumption in the UK for the period 1950 to 1976 and the associated GDP graph.

Clearly, electricity consumption is closely linked to the rate of change in real GDP (the regression equation is shown on the graph – time trends have not been eliminated given that the interest is in forecasting). How far the two are likely to be 'decoupled' depends therefore on (*a*) the extent to which non-electrical sources of energy can and will serve final uses now met by electricity (the minimum needed is the so-called 'firm' electricity for such things as lighting); and (*b*) the extent to which final uses now met by electricity are likely to be subject to 'saturation' effects (once a given amount is used for cooking, heating and lighting, it is unlikely to increase as income increases). These are the kinds of consideration that determine energy futures as far as the electricity production sector is concerned, and this was one of the factors debated at WPI.

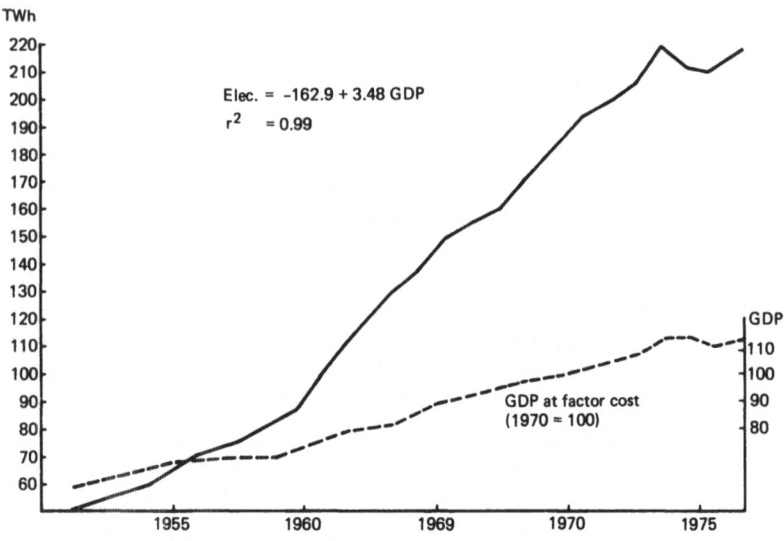

FIG. 7.2 Electricity demand and UK gross domestic product

It was perhaps a matter of good fortune that the Leach and Chapman scenarios were independently produced. The Department of Energy intends, however, to produce scenarios based on these, and other 'low-energy' futures, in official publications, without, of course, indicating that they are anything other than alternatives proposed by outside research workers. This is valuable simply because the WPI Report does not record the nature of these scenarios. Instead it takes an interesting approach to their credibility.

First, although arguments were put forward to the effect that THORP was needed to generate a plutonium inventory for fast reactors *if* it was decided to introduce them – i.e. THORP was an 'insurance policy' – Mr Justice Parker rightly pointed to the fact that certainly the first, and in his view the first eight, fast reactors could be fuelled from stocks of magnox plutonium which would have been separated anyway, whether THORP exists or not. Whatever the correct number, the fast reactor argument was not relevant *unless* fast reactor policy actually does exist, as it were, 'in the vaults' and is committed to a very rapid policy of introduction. Hence, while THORP does not imply a fast reactor programme, it is arguably a necessary condition for one to exist:

that is, the plutonium fuel must either come from reprocessing plant in the UK or be bought from such plants abroad. Given that those countries engaging in reprocessing will have their own demands for plutonium, this latter possibility seems unlikely.

Second, the Report stresses the uncertainty surrounding the alternative scenarios and forecasts. The Inspector did not feel it was his task to make his own forecasts since 'any forecast which I might make would be as uncertain as any other forecast' (Parker, 1978, para. 8.37). This is surely correct. Indeed it perhaps underestimates just how uncertain a personal forecast would have been since, in the Leach and Chapman cases at least, considerable investigative effort, on a team basis, had gone into producing their own scenarios. Given that no investigative facilities were available to the Inspector, he could not have been expected to produce independent assessments.

But if 'need' was to be established, *some* forecast had to be accepted. Mr Justice Parker's attitude was that he had no option but to rely on the official Department of Energy forecasts, not because they were necessarily any better worked out than, say, Leach's or Chapman's, but because they were produced by persons whose responsibility it would be to answer for them if they were wrong. The oddity here is quite why so much time was taken at WPI debating alternative futures based on alternative forecasts if there existed a prior view that such forecasts could not have credibility, or at least as high a credibility as the official forecasts, simply because those who made them were not in official positions. Effectively, this establishes one status for the objecting party and a quite different one for the defending party which, as an agent of Government, simply used the official forecasts. The suggestion here is not that the Inspector was wrong in attaching more weight to one set of forecasts rather than any other, but that, *if* forecasting is to be debated (as it is now to be in road inquiries) then something is wrong if the debating parties cannot begin with an *initial* position of equal credibility. If, on cross-examination, that credibility is altered, one way or the other, that would be an appropriate procedure. What seems distinctly at variance with the very concept of a public inquiry is to adopt a policy of unequal credibilities. Suitably extended it would of course mean that whatever the arguments in any inquiry they would be weighted in favour of the Government agency (if one is involved). This, of course, is exactly how some objecting parties see public inquiries, and this may partly explain the widespread tendency to describe such inquiries as 'cosmetic exercises', 'whitewashing', etc., and a regrettable but apparently increasing reluctance to participate on that account.

The sensible way out of this problem is for such national dimensions to be debated nationally. No local inquiry can provide the proper format for a reasoned assessment of the forecasts on which an investment's social desirability is partly based. These dimensions should be removed from the domain of the local inquiry and placed in a context where investigative power exists (as with Royal Commissions or Planning Inquiry Commissions or, to some extent, Select Committees of the House of Commons).

The second strand of the Inspector's argument was that energy policy had to be 'flexible' – i.e. designed in such a way that, if a given energy demand occurs, it can be met. To this end, he argued, one should invest in nuclear power to ensure that if it is needed it is there. This argument echoes official statements, as in the Energy Policy Green Paper of 1978 and elsewhere (including the official statements made by the Department of Energy to the WPI). Flexible planning is of course a laudable objective. It cannot however be left there, since the taking out of insurance costs money. That money has an opportunity cost, namely taking out other insurances. On the assumption that all insurances cannot be taken out at once, they must be ranked according to their costs and benefits. Clearly, such a ranking will favour investments where the benefits are more likely to be realised and therefore investments where considerable research and development have already taken place. This, argues the Inquiry Report, is a feature of flexible planning that supports the case for nuclear power. This is correct provided (*a*) the uncertainty attached to future nuclear installations is low compared to those for substitute electrical technologies; and (*b*) it cannot be demonstrated that diverting, say, the cost of a first fast reactor to other R&D would not result in equal or greater energy production by the time fast reactors would have come 'on line'. As far as argument (*b*) is concerned the line of reasoning would support the case for investing in further thermal reactors, since the status of alternative technology is not such that it can impact on energy outputs until the twenty-first century. This explains why a nuclear component appears in the arguments in the Leach and Chapman scenarios put to WPI. The same argument would not be legitimate for fast reactors unless quantification showed it to be so.

The salient characteristic of the Inquiry Report, however, remains the fact that, having cast doubt on the credibility of alternative energy forecasts, and having emphasised that all forecasts are uncertain (including the official ones), there is an underlying theme that higher credibility is attached to official views rather than non-official ones. Moreover, having denied that sufficient evidence was presented for a 'definitive forecast' (given that no such thing could exist, anyway), and having denied the value of any personal forecast, even if it could have been made, the Inspector implicitly accepted a specific forecast, the official one. This is illustrated by the support expressed in the Report for a programme of nuclear power and by references to some of the forecasts (that of Arthur Scargill of the Yorkshire Area Mineworkers' Union) as 'fanciful'. It is presumably not possible to know what is 'fanciful' without having a prior view of what is reasonable.

Two features therefore seem worthy of comment. First, the Report is not, in any way, a record of the debate over energy futures. Neither does it record the fact that, for many, energy futures are contingent upon rates of economic growth and this growth was itself regarded as an undesirable feature of the future. This is not to suggest that the Report should have evaluated whether economic growth is desirable or not. Somewhere in the decision-making process, however, that aspect, and it is fundamental, requires debate, as has already been noted. The misfortune is that the Report does not indicate what the debate *was* about, so that as an informational document it is less than sufficient. Just how far the planning process should (or does) require a report of proceedings is discussed later. For the moment, it can be argued that this is no criticism of the Inspector since he interpreted his function as not recording evidence that, in his view, did not assist him in making his decision.

However, attention needs to be drawn here to the Department of the Environment's 'Notes for Guidance to Inspectors' which state (para. B.2.3): 'At the same time the report should satisfy the parties to the inquiry that their submissions have been adequately and fairly reported.' This remains an unsatisfactory aspect of WPI, one which has given rise to much dissatisfaction among participants. It is desirable that in future the Inspector's responsibilities in this respect be made clear, publicly, before an inquiry begins.

The strength of this argument appears to be increased if one considers the role of the inquiry report in the overall decision-taking process. The report serves as a means of relaying to a broader public the arguments that were advanced at the inquiry, and here one must surely include many MPs and civil servants who need to be informed but cannot, understandably and properly, afford to attend the inquiry itself. In the absence of a report which gives a full account of what transpired, anybody who either did not attend or who does not possess a full set of transcripts (and that means almost the entire population) has **no way** of gaining an understanding of the inquiry.

A further reason for advocating a fuller form of report relates to the arguments presented earlier concerning the status of the objectors and the legitimacy of intelligent dissent. If the objectors' arguments are not 'adequately and fairly' reported it surely encourages objectors to infer that their arguments are taken to have little value and, by implication, that criticism and dissent are tolerated but not taken seriously. It is emphatically not suggested that this is Mr Justice Parker's or anyone else's view. It is suggested that reports like the WPI Report encourage objectors to *believe* that their submissions are so viewed and for this further reason we believe that further reports, in line with the 'Guidelines to Inspectors', would be more satisfactory in future.

The second point is that, if opposing parties cannot be seen to *begin* from positions of equal credibility then the subject-matter in question should not be dealt with in a context that generates that inequality. If, after examination of the arguments, the credibility of each party is changed (relative to each other) then that is a proper function of the exchange of views and the cross-examination process. But, in any event, the complexities of national energy policy are such that close debate on them seems inappropriately placed within the context of a local inquiry. To this end some secondary process is required whereby that debate can take place.

(c) Reprocessing versus Storage of Spent Fuels
The substance of the objectors' case to reprocessing as a *waste management option* lay in the argument that spent fuels from AGRs and any future oxide fuel reactors are best stored for some period rather than being reprocessed. That storage in 'ponds' of water could not be indefinite, no one doubted. The argument, as presented by Friends of the Earth, was that it could be so stored for a period of time which would permit a full appraisal of the two immediately obvious 'cycles' – storage and disposal or reprocessing with uranium and perhaps plutonium recovery, and disposal of the remainder. The need for such a time-period of assessment of the options was necessary, they argued, because (*a*) storage was practised elsewhere (in the USA and Canada) without evident problems; (*b*) storage avoided the proliferation problems that were entailed, in their view, by reprocessing and the radiation levels from reprocessing; and (*c*) the UK nuclear fuel industry had not given

serious consideration to fuel cycles which did not involve reprocessing. The obverse argument was that storage was possible only for a limited period; that it would necessitate the disposal of larger volumes of waste than via the reprocessing route and especially larger quantities of long-lived actinides, and that it was economic, especially if foreign contracts are allowed for. The economic arguments are considered later. For the moment it is interesting to see whether the WPI generated sufficiently clear information on the two alternatives.

Storage for *some* period is inevitable anyway since THORP cannot be 'on line' at exactly the same time as spent oxide fuels come from the AGR programme. The doubts of the industry seemed to surround stainless steel clad fuel elements which, as Chapter 6 showed, are the relevant ones for UK AGRs. No one seemed to doubt that such elements could be stored in ponds safely for ten years. The argument was in part over whether the period could be longer than that. In the event, the main force of the debate veered away from whether this was possible or not, since the storage facilities initially under consideration were the ponds – the so-called 'wet storage system'. What was argued was that dry storage in a gas medium would enable very long storage possibilities.

Uncertainty clearly surrounds the storage possibilities of spent fuel and, as Mr Justice Parker remarks, if the decision was not to reprocess, an immediate programme of research and development would have to be embarked upon to establish storage facilities and to ascertain the likely effects of the ultimate storage of spent fuel in repositories such as the sea-bed or in rock formation. The choice therefore seemed to be between a reprocessing facility for which R & D had taken place and storage facilities for which R & D had not taken place. It appeared to be generally accepted that storage avoided the proliferation problem and the radiation emissions from reprocessing. Neither did anyone doubt that the recovery of uranium and plutonium gave a 'fuel credit' to reprocessing however that fuel was used (thermal or fast reactors), but they did dispute the *size* of those credits. It was also argued that unreprocessed spent fuel had environmental hazards of its own, although the objectors regarded these hazards as minimal compared to those from reprocessing.

Since great stress was placed on plutonium separation (because it generates the proliferation risk) Mr Justice Parker placed stress on the amount of plutonium that would have to be 'finally' disposed of along either route. His argument was simple. For a given amount of waste containing, say, 100 tonnes of Pu, the storage and disposal route would mean that it would be about a quarter of a million years before the decay process reduced the Pu quantity to 0.1 tonne.[5] On the reprocessing route, only 0.1 tonne of Pu would remain in the waste that has finally to be disposed of, the bulk of the Pu and uranium having been separated out for re-use. This 0.1 tonne would over the quarter-million years decay to 100 grammes. Thus, the reprocessing route left about one-thousandth of the Pu still 'active' after a quarter-million years compared to the storage and disposal route. Or, to say the same thing, only one-thousandth of the Pu appears to be disposed of in the reprocessing route. (Actually, since the decay rate is the same, the initial values and the values after a common time-period must, expressed as a ratio, be the same, so that it is unnecessary to calculate the ratio after a quarter-million years – it is the same ratio as in the initial disposal period.)

Of course, what has been forgotten in this comparison is the fact that the separated plutonium and uranium must themselves have an ultimate destination and must themselves be treated. Since reprocessing actually *adds* to waste streams (see below), it is not necessarily the case that *less* waste has to be disposed of via the reprocessing route than via the storage and disposal route. The objectors drew attention to this, although it seems fair to say that their case has been more cogently presented after the Inquiry than during the Inquiry (see Conroy, 1978 – this text was published before the Inquiry Report was published but after the Inquiry itself was over). Essentially, no UK figures appear to exist for the comparison of the volume of spent fuel to be disposed of compared to the total wastes from reprocessing that would have to be disposed of. Conroy quotes what he considers to be an upper limit figure of 30 cubic metres (a 'normalised' figure to make direct comparison easier – see Conroy, 1978, pp. 15–16) for a 600 tonne per year plant. The comparable figures for the reprocessing route were given by BNFL at the WPI. These are 32.8m³ for compacted waste and 47.5m³ for uncompacted waste. Thus, if it is legitimate to use the US figure, reprocessing involves the disposal of *more* waste, not less than the storage and disposal route. Conroy quotes other US sources as suggesting that the reprocessing route also generates wastes with higher heat content and hence wastes which need larger amounts of space in which to be stored.

Whatever the truth about this comparison, it is understandable that the emphasis placed on the disposal of plutonium should have led the Inspector to think mainly in terms of how much of this one substance would be disposed of by each route. Obviously, if the separated plutonium must itself be disposed of then the amount of plutonium is not reduced at all. But since it is unclear why a reprocessing plant would exist if what was separated was simply stored, Mr Justice Parker rightly assumes that it will be used as a fuel. The next subsection of this chapter, on the economics of THORP, indicates how the prospective financial return achieved by THORP is affected by assuming the plutonium is used as a mixed fuel in thermal reactors and how it is affected by assuming it is used in fast reactors. It seems possible that the fast reactor option could not be stressed in the Report for fear that it would imply some commitment to at least one fast reactor, a commitment which cannot be made until the inquiry into CDFR takes place.

In his Report, Mr Justice Parker illustrated the plutonium inventory argument by reference to use of the Pu as a fuel in thermal reactors. He assumes a 10 GWe per annum AGR capacity available to use the plutonium as a fuel. The AGRs' spent fuel would at the end of a given period be 12 tonnes, to which is added 55 tonnes of magnox plutonium. The latter is separated plutonium so that the inventory reads: 55 te.Pu (separated) +12 te.Pu (unseparated). But if the AGRs use Pu as a fuel they can take an estimated 46 tonnes. Of this, 32 tonnes would 'reappear' in the spent fuel. Hence the inventory is now 32 te.Pu (unseparated) +9 te.Pu (55 −46) (separated). The total stocks are therefore 67 by the storage route and 41 by the reprocessing route, a reduction of 26 tonnes.

Mr Justice Parker notes that 'the price of the [above] reduction is of course the discharge of radioactivity involved in the reprocessing' (para. 8.26). If Conroy is right, however, the price is higher than this. He points out that plutonium-fuelled reactors have greater concentrations of actinides, particularly americium and curium. Allowing for the fact that these also decay to plutonium 239 and 240, Conroy calculates the total plutonium content for the two routes and arrives at a ratio of about 7 to 1 in favour of the reprocessing route. This, he argues, should be compared to the BNFL figure of 1000 to 1. The 'plutonium inheritance' argument stays in favour of reprocessing but not on anything like the magnitude suggested.

As far as the WPI Report is concerned the analysis is taken no further than the hypothetical AGR example cited above. The 1000 to 1 ratio is, however, firmly stated and is not adjusted after the calculations on the use of Pu as a thermal reactor fuel.

Since the Inquiry could not effectively deal with debates about fast reactor cycles, the disposal of plutonium to a fast reactor was not considered in great detail. What was required was full fuel cycle evaluations to assess the impacts on waste disposal. It is now the case that advocates of fast reactors are pointing to their potential for 'incinerating' plutonium – i.e. not setting the reactor to 'breed' but to produce *less* plutonium than it consumes. How credible this is as a scenario depends on one's belief about the future state of the uranium market and therefore, by implication, the implied value of plutonium as a fuel. Deliberately to incinerate a valuable fuel would seem odd if, at one and the same time, one is arguing that fast reactors are needed to overcome a uranium supply shortage. Of course, an ultimate 'balance' between thermal and fast reactors could be achieved so that the separated plutonium from thermal reactors is just sufficient to 'feed' fast reactors which can be 'tuned' so as to make inputs and outputs of plutonium just equal. It is unclear, however, if that is an option to be countenanced much before the middle of the next century. If so, the 'incineration' hypothesis has little bearing on the plutonium inventory argument since it would seem more sensible to store the plutonium for use as a fuel.

Finally, the reprocessing route requires the vitrification of the highly active component of waste, unless other technologies emerge, and the final disposal of that waste. Vitrification has existed since the 1950s as a proven laboratory technology in the UK. It does not exist on a commercial scale and BNFL have been given permission to construct a commercial-scale plant. Since highly active waste already exists and since more will come from magnox reprocessing, vitrification or some other process is essential regardless of THORP. This is stated clearly in the Inquiry Report (para. 8.30). It may seem odd, however, to have embarked on a programme of nuclear power before demonstrating vitrification and the acceptability of final disposal on a scale larger than has so far been achieved. The general argument explaining this is that, to date, the quantities of highly active waste in the UK have not been sufficient to develop a technology which the industry is confident merely requires 'scaling-up'.

(d) The Economics of THORP

In most public inquiries the economic viability of the plant in question is not discussed. This seems legitimate where the appellant is a private company or individual. It is less reasonable where the party is a nationalised concern or public agency or subsidiary thereof. It could be argued that such concerns are already subject to adequate financial scrutiny through such bodies as the House of Commons Public Accounts Committee. In effect, however, that committee can only assess projects that are completed or at least well under way. In 1978, for example, the Public Accounts Committee criticised the cost overrun on the Humber Bridge. In the past, Select Committees have requested financial appraisals of investments in order to satisfy themselves that the 'right' financial decision is being made. That such scrutiny can readily go astray is amply documented in the case of the choice of the AGR over the LWR – see Burn (1978).

The public inquiry, whether local or PIC or in the form of a national commission, affords a singular chance for detailed scrutiny of planned investments by bodies that should in any event be publicly accountable. Interestingly, in law, there is no requirement on them to prove financial viability as part of their case, nor even to present a cost-benefit style proof of evidence. Given that the latter has already occasioned extensive comment, especially in terms of its use at the Third London Airport inquiry in 1969–70, and given that it has special problems when attempts are made to apply it to energy investments (see Pearce, 1979), it seems reasonable perhaps to expect social cost-benefit studies to be presented but to be supplemented by other 'impact' statements. Detailed financial appraisals should be presented. Where financial appraisals reflect the profitability assessment by the investing authority in question, cost-benefit analysis and impact statements at least attempt to indicate the returns and costs to *society* as a whole.

Such appraisals should form part of the 'requirements' of an efficient inquiry procedure. This need not entail that financial viability should be a condition for granting planning permission. In this Mr Justice Parker was quite correct to note that factors other than financial viability will be relevant to such a decision (Parker, 1978, para. 9.1). It is difficult, however, to see why, simply because the law is what it is, it should be permitted to dictate procedure when it is evidently desirable that public accountability be served by having detailed financial analysis presented. Moreover, that analysis should be based on sound accounting procedures. This is hardly different from making a requirement that, say, an environmental impact statement should take on some specific form or that an Inspector's report should have a specific format, and so on.

In the event, financial appraisals were made of THORP. Mr Justice Parker was critical of the evidence presented by BNFL to this effect and little has changed since then. He was therefore obliged to carry out his own elementary financial analysis using BNFL data and the submissions of Dr Peter Chapman. A post-inquiry analysis can be found in Conroy (1978) which updates Chapman's analysis. Because none of these is totally satisfactory, this section presents a separate *outline* attempt to evaluate the economics of THORP. It does so not because of any desire to evaluate the arguments presented, but to indicate the *type* of analysis one would expect to see in any future inquiry. The emphasis is therefore on the methodology rather than the absolute magnitudes. For more detail see Pratt (1978).[6]

The essential requirements for a financial appraisal are:

(i) that it should be presented in discounted cash flow terms, expressing results in terms of a net present value or internal rate of return (or both); (ii) that it should contain extensive sensitivity analysis to show the results of varying the assumptions about, e.g. fuel prices, on the outcome.

An attempt is made to give an overview of such an approach here. Much detail has been omitted and can be found in Pratt (1978).

THORP is designed to extract uranium and plutonium from spent oxide fuel. If THORP did not exist some other plant would have to exist to store and eventually dispose of waste. Hence a proper comparison requires a calculation of the difference between the net present values of the two routes. In doing this we assume that THORP operates to capacity, an assumption which embraces the official forecasts noted earlier, and that foreign contracts are sought in such a way as to maximise the use of capacity. Various conventions have had to be used because of the paucity of data, especially as BNFL had not evaluated the storage/disposal route in any detail. One other procedural point is very important. During the WPI it would seem that 'advice' was tendered to BNFL not to justify THORP as a plant designed to inventory and 'sell' plutonium for fast reactors. The rationale for this appeared to be that, ostensibly, the United Kingdom is not even committed to *one* commercial-scale fast reactor, let alone a programme. To have presented the economics in terms of THORP's capacity to supply plutonium to fast reactors would, on this argument, have been to prejudice the inquiry into CDFR. It is unclear that any such consequence would have followed, however. In what follows we show the results of assuming that the recovered uranium and plutonium is used in thermal reactors only, and assuming that the plutonium is used in fast reactors. Since it remains the view of many that THORP is a symbolic commitment to fast reactors (since its justification is stronger, allegedly, if it produces Pu for fast reactors) this issue seems worth investigating.

Estimates of spent fuel arisings from AGRs are available from BNFL documentation at WPI. These arisings assume an AGR programme that is not perhaps the same as that envisaged in the Department of Energy's

Green Paper, but the results are not greatly affected. To these arisings are added spent fuels received from abroad on foreign contract. The cost of transporting this spent fuel to THORP is identical for the reprocessing route or the storage/disposal route. The costs are shown in column 1 of Table 7.2. To these costs must be added storage costs for the reprocessing route and these are shown in column 2. Reprocessing operating costs are shown in column 3 and are based on the assumption that by 1997 THORP will have processed 6000 tonnes at an estimated total cost of £140m. The figures vary from year to year because of the different tonnages of spent fuel being processed in any one year. Note further that foreign fuel is not now included in the calculations. This is because it is easier to add in profit per tonne of reprocessed foreign fuel at a later stage. Hence, of the 6000 tonnes, it is estimated that 3150 tonnes will be of UK origin and 2850 tonnes of foreign origin. Thus all the columns so far relate to UK fuel only.

In the reprocessing route the next stage will consist of the fabrication of the uranium and plutonium for re-use. To secure these costs we consider the amount of spent fuel and the amount of U and Pu that can be recovered from it to get an equivalent in terms of enriched uranium. These totals can then be multiplied by unit costs of fabrication to get the fabrication costs in column 4. To get capital expenditure we take a fraction of the total capital expenditure for the whole plant (which deals with UK and foreign fuels) and in the absence of better guidelines simply use 0.525 multiplied by the total estimated capital cost (i.e. $0.525 = 3150/6000$) and allocate it across years according to the capital expenditure plans of BNFL. This gives column 6.

Now, whether there is a reprocessing route or a storage/disposal route there must be ultimate disposal in stable rock formation or elsewhere. We have already noted that a vitrification plant is to be built. The convention followed here is that, given the even greater uncertainty of the costs of disposing of unreprocessed waste, the two final disposal costs are regarded as being approximately equal. The caveats here must be that the reprocessing route may, as we have seen, involve more and not less waste to dispose of and that vitrification is not needed until later in the storage route. None the less, it is not clear if these extra costs are not offset by differences in the levels of treatment before disposal. As a working rule, therefore, we assume equal final disposal costs and, since they will 'net out', we leave them out of the analysis. This is unsatisfactory, but appears to be the best that can be done for the moment, given the poor state of the data.

So the total of reprocessing costs is given in columns 7 and 8. Column

TABLE 7.2   Costs of reprocessing versus storage: THORP (£m, 1977 prices)

			REPROCESSING									STORAGE	
Year	Transport Bottles	Storage Operating Cost	Reprocessing Operating Cost	Fabrication Cost: U	Fabrication Cost: Pu	Capital Expenditure	Total Cost U only	Total Cost U+Pu	Foreign Profit	Net Cost (U only)	Net Cost (U+Pu)	Total (U only)	Total (U+Pu)
	1	*2*	*3*	*4*	*5*	*6*	*7*	*8*	*9*	*10*	*11*	*12*	*13*
75/6	0.31						0.31	0.31		0.31	0.31	0.31	0.31
76/7	0.19						0.19	0.19		0.19	0.19	0.19	0.19
77/8	0.78					16.07	16.85	16.85		16.85	16.85	3.71	3.71
78/9	0.97					16.07	17.04	17.04		17.04	17.04	9.76	9.76
79/0	1.38					16.07	17.45	17.45		17.45	17.45	18.96	18.96
80/1	1.72	0.68				16.07	18.47	18.47		18.49	18.47	25.16	25.16
81/2	1.80	0.62				16.07	18.49	18.49		18.49	18.49	30.37	30.37
82/3	1.90	1.24				39.69	42.83	42.83		42.83	42.83	25.34	25.34
83/4	2.05	1.22				39.69	42.96	42.96		42.96	42.96	14.51	14.51
84/5	2.05	1.86				39.69	43.60	43.60		43.60	43.60	15.24	15.24
85/6	2.05	1.86				39.69	43.60	43.60		43.60	43.60	15.24	15.24
86/7	2.05	1.77				39.69	43.51	43.51		43.51	43.51	19.54	22.92
87/8	1.00	1.71	2.45			5.15	10.31	10.31	2.99	7.32	7.32	25.16	31.93
88/9		1.71	4.90	6.12	3.75	5.15	17.88	21.63	5.98	11.90	15.65	43.84	59.00
89/0		1.74	8.28	12.25	7.51	5.15	27.42	39.67	10.08	17.34	29.59	50.81	69.68
90/1		1.80	8.28	20.71	12.69	5.15	35.94	48.63	10.08	25.86	38.55	64.10	88.10
91/2		1.92	8.28	20.71	12.69		36.06	48.75	10.08	25.98	38.67	67.76	91.76
92/3		2.13	8.28	20.71	12.69		31.12	43.81	10.08	21.04	33.73	67.76	91.76
93/4		2.61	8.28	20.71	12.69		31.60	44.29	10.08	21.52	34.21	67.76	91.76
94/5		2.16	8.28	20.71	12.69		31.15	43.84	10.08	21.07	33.76	67.58	91.48
95/6		2.55	8.21	20.71	12.69		31.47	44.16	10.08	21.39	34.08	63.98	87.91
96/7		3.00	8.24	20.53	12.59		31.77	44.36	10.24	21.53	34.28	42.44	54.90
97/8		(?)		20.59	12.62		20.59	33.21		20.59	33.12	32.95	45.44

All data based on BNFL submissions to WPI.

7 assumes a 'uranium only' recovery, so it is equal to columns 1–4 plus column 6. It excludes the plutonium fabrication cost, column 5. To get column 8, in which plutonium and uranium are *both* recovered, add columns 1 to 6. To offset such costs there are the profits on foreign contracts. Little is known of these contracts because they are commercial secrets. None the less, the Japanese contract price is known and it is also known to be on a 'cost-plus' basis. Therefore, by subtracting the relevant costs one can arrive at a figure for the profit of some £31,500 per tonne (Pratt, 1978).[7] Column 9 gives the profit figures and columns 10 and 11 show the resulting net costs for the U and U + Pu routes.

Exactly the same procedure is carried out for the storage route. The essential points are that there will be transport costs, storage operating costs and capital expenditure on storage plant. The last is less well known than with reprocessing since BNFL had given little attention to the cost of such facilities. Significantly, however, the storage route does not result in the recovery of U and Pu so that uranium must be bought in amounts equal to the fuel equivalent of the recovered fuels from reprocessing. Otherwise like is not being compared with like. The amounts of recovered fuels have already been calculated in working out the fabrication costs and hence it is necessary to buy 'yellow-cake' (uranium in its imported state) in an amount sufficient to generate these quantities. Generally, 1 tonne of fuel requires 6 tonnes of uranium ore and this in turn costs some $30 per lb to buy (uranium prices are themselves complex to deal with – for more detail see Pratt, 1978). This ore must be fabricated and enriched. When all these are added together we secure columns 12 and 13 of Table 7.2. Column 12 covers the use of uranium only and column 13 covers the purchase of uranium sufficient to equal the U + Pu route in the reprocessing option.

Thus, if column 13 is deducted from column 11 we have the net cost of reprocessing with U and Pu recovery. We see that this cost quickly becomes negative, indicating that THORP is *superior* to storage and disposal.

However, it is necessary to apply a discount rate to the annual data to get it back to present value terms. If a 10 per cent rate is used we secure a net present value advantage to THORP of £19.2m if uranium only is recovered and £42.6m if both uranium and plutonium are recovered. Notice that these are *cost advantages to THORP*.

Confining ourselves to this cost advantage for the moment it is instructive to see what would have to change for this advantage to disappear. If the discount rate is lowered to 7 per cent the present value cost advantage rises to £52.5m and £87.5m respectively for the U and

U + Pu routes. The information on cost escalation is more interesting. Assuming storage costs were not themselves subject to the same cost escalation, the cost of THORP would have to rise in terms of capital expenditure by only 14 per cent to make the uranium recovery route unfavourable compared to storage and only 30 per cent to make the combined recovery route unfavourable. We say 'only' because such cost overruns are not the exception, but the rule, in nuclear facility construction (Burn, 1978). Of course, the chances that the storage route would experience similar cost overruns are high, especially as long-term storage of stainless steel clad elements seems to be a technology that little is known about. Then again, the same remark is true of large-scale reprocessing.

Note too that the reprocessing route has two further technologies associated with it which are not relevant to the storage route. First, a krypton removal plant has to exist (this is a recommendation in Parker's report and a requirement laid down by the Government). This may not, however, be a significant cost assuming such a plant can be perfected. More important, considerable talk has taken place about the conditions under which recovered fuels would be returned to countries of origin. The usual answer is that they would be 'spiked' – i.e. irradiated in such a way that they could only be used in the reactors in the country of origin. No one it appears, seems to know the cost of spiking. On one argument it doesn't matter how expensive it is since the foreign contracts are on a cost-plus basis. This is hardly tenable as a view, however, since if it is very expensive it must affect the terms of the contract unless countries sending spent fuel to THORP behave in a totally price-inelastic fashion. This hardly seems credible. On some arguments spiking could be extremely expensive, especially as different 'formulae' may be needed for different countries of origin. No one can pretend that this issue was satisfactorily settled at WPI. The costs of spiking are not mentioned in Parker's own financial appraisal and it is quite unclear if spiking costs are incorporated in contract prices.

Finally on 'sensitivity' analysis, one would need to know what would happen if uranium prices change. On Pratt's analysis (Pratt, 1978) a 1 per cent rise in the price of uranium would raise the NPV of THORP over storage by about $2\frac{1}{2}$ per cent. The NPV is not quite so sensitive to the U + Pu option but this probably reflects the time-period considered. Certainly, the price would have to *fall* to under $17 per lb for THORP to cease to have a cost advantage over storage.

Now, if the choice was between storage and reprocessing simply because *something* has to be done with spent fuel then the preceding analysis would seem to dictate that the economic arguments would favour THORP. An interesting question in whether THORP is itself profitable, since the alleged commercial gain to the country was used at WPI to favour investment in THORP. This issue can be briefly considered. First the amount of plutonium extractable from spent fuel is calculated. Second, a hypothetical fast reactor programme is postulated. For the calculations here it is assumed that a second FBR comes on line in 1995 and a third in 2000. Thereafter one 1.3 GW reactor is commissioned each year. Allowing for the initial charges and the annual charges needed for these reactors, and for the fact that existing stocks of Pu are likely to be used first, we can build up a picture of flows of Pu going into fast reactors. The existing stocks of magnox plutonium are exhausted by 2002 and THORP plutonium is used thereafter. Interestingly, and in keeping with other comment, such a scenario requires a *second* THORP at the turn of the century. But this is incidental to the main analysis. By comparing the efficiency of plutonium and uranium and attaching a shadow price to the plutonium based on the 'market' price of uranium and allowing for the different quantities needed to get the same fuel equivalent, one ends up with a net credit of some £27m in present value terms for the reprocessing route with plutonium recovery. That is, the plant's net *cost* advantage over storage is increased from about £43m for U + Pu recovery to about £70m.

(e) Proliferation
An appreciation of the significance of the submissions made to WPI on this topic is helped by a brief account of the development of the global context at that time.

It has been feared from the earliest days that the development of nuclear power would lead to the proliferation of nuclear weapons. However, civil nuclear power growth took place in the belief that proliferation could be avoided by three major deterrents:

(i) The technological complexity of weapons manufacture, which would deter most states.

(ii) The difficulty of obtaining nuclear explosive materials – plutonium or highly enriched uranium. It was widely believed that the plutonium produced by nuclear power reactors could not be used to manufacture bombs.

(iii) The inspection activities of the International Atomic Energy Agency (IAEA) set up in 1956.

Thus the IAEA statute (1956), the Euratom Treaty (1957) and the Non-Proliferation Treaty (1970) all reflect the belief that fast reactors, reprocessing facilities and the transport and stockpiling of plutonium could all develop on a global scale without significant risk of the proliferation of nuclear weapons.

By the time of the WPI, however, the position had changed and the above deterrents were seriously in question. First, technological complexity was no longer believed to be an obstacle to even a technologically relatively unsophisticated state or, even, sub-national groups securing a weapon made from civil fuels (Barnaby, 1978). Secondly, it was known that reactor-grade plutonium could be used to produce a credible nuclear weapon, as Professor A. Wohlstetter proved in evidence to WPI. Third, the IAEA's capacity to detect the acquisition of nuclear explosive materials was widely criticised (Johnson, 1977). Moreover, given that plutonium can be inserted into a previously prepared device in a matter of hours or days, even if acquisition is detected, this does not offer *timely* warning, because the response-time of the international diplomatic system, which is the only other deterrent, is to be measured in weeks or months rather than hours or days (Nye, 1978).

This meant that the exporting of fast reactors, plutonium or highly enriched uranium, or the sale of reprocessing or enrichment facilities to non-nuclear-weapons states, were *all* potentially dangerous activities, because any of these activities could lead to the rapid completion of nuclear weapon fabrication by the recipient state. Growing awareness of these problems led to legislation in the USA – the Nuclear Non-Proliferation Act (1978). The policy in this Act may be seen as an attempt to use the fact that the USA controls about 90 per cent of the non-Communist states' supply of reactor fuel as a bargaining counter. The idea is that a guaranteed fuel supply and spent fuel storage facility, possibly under international control, could be traded for the promise to forgo national development of enrichment and reprocessing technologies and fast reactors (Nye, 1978).

This policy, which forms the background to WPI, was summarised by Mr Justice Parker in his Report. For convenience, we reproduce his précis of it here:

The policy was developed by President Carter at a news conference on 7 April 1977. It comprised in essence the following:
1. Indefinite deferment of commercial reprocessing and recycling of plutonium.

2. Giving increased priority to the search for alternative designs for the FBR and deferring the date when FBRs would be put into use.

3. Increasing US capacity to provide adequate and timely supplies of nuclear fuels to countries that needed them *so that they will not be required or encouraged to reprocess their own materials.*

4. Proposing to Congress the necessary legislation to sign supply contracts and *remove the pressure for the reprocessing of nuclear fuels by other countries which did not then have that capability.*

5. An embargo on the export of equipment or technology that could permit uranium enrichment or chemical reprocessing.

6. Pursuing discussions of a wide range of international approaches and frameworks that would permit all countries to achieve their own energy needs while at the same time reducing the spread of the capabilities for nuclear explosive development.

Under the last heading the President mentioned the establishment of an International Fuel Cycle Evaluation Programme (INFCEP) 'so that we can share with countries which have to reprocess nuclear fuel the responsibility for curtailing the ability for the development of explosives.' The INFCEP has since then been established. The President also mentioned that the US would have to help to provide some means for the storage of spent fuel and, since that time, plans have been announced for the US to receive and store such fuel. (Parker, 1978, para. 6.21. Mr Justice Parker's emphasis has been left in order to indicate his own assessment of the important wording.)

The main burden of several submissions to WPI was that THORP should be rejected because of its proliferation risks, several among them echoing the RCEP 6th Report (para. 506) saying' . . . the action we take in response to our assessment of these risks could have a substantial impact on world opinion.' The basis of these submissions was that, along with Japan, France and West Germany, the UK should renounce reprocessing because that would lend support to the USA's position on proliferation and, conversely, that a decision to reprocess by the second-oldest nuclear weapons state would 'legitimise' the activity and make it very difficult to persuade others to refrain.

Mr Justice Parker's Report notes, but does not comment on the fact, that nuclear explosive materials have been exported by the USA and the UK under safeguards, and that these safeguards 'could and should be improved' (para. 6.6). Turning to the statutes governing IAEA, Euratom and the NPT, he observes that they were drawn up at a time when the export of reactor-grade plutonium was regarded as acceptable and interprets the NPT to mean that nuclear-weapons states are *obliged* to furnish reprocessing facilities to non-nuclear-weapons states. Mr Justice Parker then argues (para. 6.24) that the denial of reprocessing facilities would encourage the development of indigenous facilities and that this effect is to be set against the proliferation effect of returning plutonium to non-nuclear-weapons states. He says that he cannot assess the relative magnitudes of these effects (6.33), but adds that the second could be mitigated by brief irradiation of plutonium-bearing fuel rods, a process known as 'spiking'. As far as the proliferation aspects of THORP are concerned, then, Mr Justice Parker argued:

(i) any resulting risk will not occur for ten years;
(ii) to refuse to supply reprocessing facilities would be against the spirit and perhaps the letter of the NPT;
(iii) the denial of such facilities would encourage the development of indigenous facilities in other countries;
(iv) there is a world need for reprocessing facilities.

The main arguments at WPI *against* THORP on the proliferation argument were several. It was argued that attention needed to be paid to the relationship between the UK and USA and international policy in the area and that this relationship did *not* support the establishment of reprocessing facilities. Further, the NPT does not lend support to the idea that the UK is, in any way, *obliged* to reprocess foreign spent fuel.

These criticisms raise several general points about the organisation of WPI. The complexity and breadth of the subject-matter must pose considerable difficulties for any one person. It seems less than fair to ask a single inspector, even with the help of expert assessors, to have a thorough understanding of such a broad range of complex issues. It is in the light of this that a Commission, with wider ranges of expertise, could assist. The use of a Commission would have a further advantage. The proliferation issue, like others raised at WPI, is capable of more than one interpretation. Since a function of inquiries is to offer advice to decision-makers, it seems reasonable to make provision for minority reports, reflecting differences of view and interpretation. For example, the interpretation of the NPT in the WPI Report suggests that the *principal* question asked was, in effect, 'can the NPT be read in such a way as to justify a UK reprocessing facility?' It is possible that someone else would have defined the problem differently, perhaps giving rise to a different analysis of the role of the NPT in current non-proliferation strategies.

There is for example the argument that the intention of the NPT was to inhibit proliferation and that, whatever its internal inconsistencies and precise legal meaning, the Treaty commits the UK to refrain from exporting nuclear explosive materials unless it can demonstrate a highly proliferation-resistent way of doing this, which, the argument runs, has not been done. A straightforward refusal to return plutonium, or any other material, to the country of origin would almost certainly produce a refusal to send it in the first place. Return 'when required for civil reactors' (WPI Report, para. 6.32) seems open to all kinds of manipulation – why, for example, should the recipient country not divert a small part of the consignment, since the critical mass of plutonium required for a bomb is less than ten kilogrammes?

There is the further question of the relationship between UK and US policy on this issue. What the US President said at his news conference of 7 April 1977 was that 'Germany, Japan and others . . . have a perfect right to go ahead and continue with their own reprocessing efforts. But we hope that they will join us in eliminating in the future additional countries which might have had this capacity evolved'. But to acknowledge a right is not necessarily to approve its exercise and it is known that Presidential adviser Joseph S. Nye sent a letter to the British Government 'which explicitly says that the US does not support the development of reprocessing plants by nuclear weapons states' (Hansard, 22 March 1978, col. 1595 – see Appendix 7 for the text of this letter). The House of Commons was assured by Leo Abse, MP that Mr Justice Parker had seen this letter at the time he wrote his Report. There was also a letter from Senator Glenn, Chairman of the appropriate Senate Sub-Committee, to President Carter, of which the Government possesses a copy, which says, referring to Windscale and La Hague reprocessing plants, 'to grant reprocessing approvals for fuel of US origin on a scale necessary for keeping the enterprise alive would be to make a mockery of our policy'. This point was strenuously advanced at WPI – see, for example, the closing submission by R. Kidwell, QC, for Friends of the Earth (Day 93 of the proceedings).

A second difficulty facing WPI concerns the question of policy. The decision concerning THORP is intimately linked with national policy on reprocessing and proliferation, and demands an understanding of that policy and the policy context and background thinking in which it is set. By the same token, it requires a grasp of the policies of other countries and institutions and the articulation of those policies with those of the UK. But the current official view is that Government policy should not be questioned at local inquiries. It is not easy to see how the various possible courses of action can be evaluated without examining the nation's policy. To claim that national policy is 'frozen', as was the case at WPI, does not seem to offer adequate scope to parties to the inquiry unless they can discuss such national issues elsewhere. Clearly, there are limits to acceptable disclosure of Government policy in such a sensitive area, but, equally clearly, the inquiry would have benefited from a more detailed account of the Government's position and the thinking behind it. This also raises the issue of the relationship between Parliament and inquiries, which is dealt with in Chapters 4 and 9.

The NPT is the outcome of long and detailed negotiations and bargaining between countries with varying interests and perspectives i.e. the safeguards 'system' is a series of negotiated political compromises. As such it is, not surprisingly, open to criticism, but there is room for disagreement about how and where improvements could be made. Whilst recognising the need for inspection, states are understandably reluctant to cede much sovereignty in this defence-related area. The force of Mr Justice Parker's remarks to the effect that the NPT 'could and should be improved' (WPI Report, para. 6.6) must therefore be tempered by an evaluation of the feasibility of such a suggestion, for it is on the question of how rigorous inspection can be made that much of the argument turns.

The point here is not that one view is 'right' and another 'wrong', but rather that several views can be sustained. There is room for disagreement as to their relative merits and these differences deserve attention at inquiries and in the subsequent reports, because they properly form a part of the advice which Parliament seeks. It is at the decision stage and not at the advisory stage in the inquiry process that a judgement about interpretation has to be made.

The use of a Commission for these issues would allow, indeed encourage, the exploration of a variety of views both at inquiries and in subsequent reports, and thereby improve the quality of the advice which Parliament receives.

Mr Justice Parker's argument that to refuse reprocessing facilities encourages the development of indigenous facilities can be opposed by the argument that, whilst this may be so, it is also possible that states might be deterred from taking this expensive step by international pressure, linked to fuel supply and disposal facility 'guarantees.' It was argued at WPI that the building of clandestine facilities might be deterred by these means and by the fact that international surveillance systems are now thought capable of detecting even fairly small facilities. The strengths and weaknesses of these arguments were presented fully at WPI but are not thoroughly explored in the WPI Report.

The idea that THORP is consistent with the relevant international treaties rests on the assertion that the treaties (the Euratom Treaty and the NPT) currently governing the production and transfer of 'special fissionable materials' – plutonium and highly enriched uranium – implicitly accept that these materials will be manufactured and supplied to non-nuclear-weapons states. Considerable sums have been spent in the belief that these expectations will be realised. But the wording of these treaties, it was argued at WPI, reflects the currency, *at the time the treaties were prepared*, of the belief that 'reactor-grade' plutonium was not suitable for bomb manufacture. Since this is not now generally believed to be true, the question must be: how, given current knowledge, can the proliferation of nuclear weapons be avoided? Mr Justice Parker's view was threefold. First, the refusal of supply encourages the proliferation of national facilities, and therefore reprocessing should proceed but be restricted to a few centres. (The contrary view is that a UK plant will 'legitimise' the activity and encourage others.) Second, inspection arrangements can and should be improved. (The contrary view is that the establishment of INFCEP took place precisely because such improvements are not known to be possible.) Thirdly, the return in the form of briefly irradiated or 'spiked' fuel rods 'might mitigate' the proliferation risk. Little mention has been made of such technical 'fixes' since WPI. The remarks that have been made have mainly referred to the obstacles to and shortcomings of the proposal, such as the need to construct additional fuel fabrication facilities and a reactor capable of accepting a variety of fuel elements, the need for remote-handling facilities at UK and customer plants, the fact that this would necessitate the disclosure to the UK of customers' fuel fabrication design details, that such brief irradiation would decay rapidly and that, should spiking fail for any reason, the element of timely warning is lost. Further, such a procedure would have to be paid for somehow and there must be a limit to what BNFL's customers will pay for reprocessing unless their demand for returned fuel is totally inelastic. If these criticisms have any force, the position taken in the Report seems to be seriously weakened, because the proliferation consequences of the reprocessing and return of customers' materials seem unavoidable unless some other solution can be found. It would have been helpful, not least to Parliament, if these aspects of the proposal had been more fully explored in the WPI Report.

Overall, then, it is clear that major international treaties were under scrutiny at WPI. The force of the objectors' case was that these treaties reflected technical beliefs at the time about the military possibilities of using 'civilian' plutonium . Further, the most recent initiative, by the USA, was held *not* to be consistent with the construction of THORP. Justice Parker's interpretation was different in that, while accepting that proliferation risk exists, he felt the USA initiative to be consistent with an actual obligation to reprocess foreign fuels. Views also clearly diverged on what the effect of THORP would be – to reduce the desires of foreign countries to establish their own plants or actually to encourage them by establishing reprocessing as an 'acceptable' part of the nuclear fuel cycle. Whichever side is correct, it is questionable if a local public inquiry is at all the proper venue for a crucial interpretation of international treaties and their intent. This venue is surely at the national level since such treaties are, in the terms we have used, relevant to the 'national agenda'.

(f) Civil Liberties
The debate about civil liberties at WPI foundered on the paradox that to discuss security measures in public threatens to reduce their effectiveness. The Inspector pointed out that Rule 10(4) of the Procedural Rules for Inquiry Inspectors says 'the appointed person shall not require or permit the giving or production of any evidence, whether written or oral, which would be contrary to the public interest'; and the Report says (WPI Report, para. 7.1):

> It would, for example, be neither in their [objectors] interests nor the public interest . . . if they were to secure the disclosure of information which would or might create or increase the nuclear capability of others . . . or expose existing or possible future installations in this country to vulnerability from terrorists.

Yet there are genuine fears. For example there is general agreement that the production, storage and transportation of plutonium, or highly enriched uranium or similar materials, will necessitate greater security precautions than would otherwise be necessary, because such activities inherently generate opportunities for diversion and hence the need for greater security activity. Secondly, it is feared that there could be

> an insidious growth in surveillance in response to a growing threat, as the amount of plutonium in existence and familiarity with its

properties increases; and the possibility that a single serious incident in the future might bring a realisation of the need to increase security measures and surveillance to a degree that would be regarded as wholly unacceptable, but which could not then be avoided because of the extent of our dependence on plutonium for energy supplies. (RCEP 6th Report, para. 332)

To be fair, the RCEP Report was speaking of a time when fast reactors are widely used, and Mr Justice Parker pointed out (WPI Report, para. 7.15) that a commitment to THORP is not a commitment to a fast reactor programme. But this serves to illustrate, it could be argued, the above point about 'an insidious growth in surveillance', for it could also be argued of the first (second, third, etc?) fast reactor that they did not represent a commitment to a fast reactor *programme*. It seems credible that such a scenario could produce exactly the situation foreseen by the Royal Commission.

Thus, whilst no one would deny the force of the Inspector's argument, it could clearly be used to defer public discussion of the problem indefinitely. It can be argued against this that any public discussion of security matters is potentially harmful, therefore public inquiries can serve no useful purpose in this sphere other than as platforms for general statements of concern. This latter view is well established in Britain, where even Parliament is effectively prevented from discussing the activities of the security services on the floor of the House or in Committee, and they may not be discussed even by the full Cabinet.

But *not* discussing the role of the security services, at inquiries or elsewhere, also has a cost. The public is aware of the existence of the Special Branch, of activities such as phone-tapping, mail interception and electronic surveillance, and of the existence of the security services' computer-stored data-banks. (The forthcoming report of the Data Protection Committee should be relevant here.) However necessary these activities and facilities may be, they inevitably carry to risk that they themselves will generate distrust, suspicion and a measure of hostility, thereby increasing the risk of subversion and terrorism, which in turn generates demands for more 'security'. 'Justice', in its evidence to WPI, drew attention to this unhappy but all-too familiar spiral, referring to International Commission of Jurists' reports on countries such as South Africa, Spain, Iran, and Chile.

Blanket secrecy about the role of the security services and arrangements for the monitoring of their activities must encourage the increasingly popular 'conspiracy' view of their functions – see for example Bunyan (1976). The same theme is taken up by Andrew (1977), who argues that secrecy 'encourages the view of the intelligence services as a sinister conspiracy designed to undermine rather than to preserve civil liberties'. He concludes: 'The only alternative to a public informed about the intelligence services is a public misinformed and suspicious of them.'

It would seem, therefore, that the paradox on which the National Council for Civil Liberties' submission to WPI focused, and which Mr Justice Parker accepted, that a discussion of security measures can impair their effectiveness, has to be balanced against the further paradox that *not* discussing security measures can itself lead to a need for increased security activity. Other arguments have also been put in favour of such discussion. First, the security services need general public support for their activities if security services are to exist – and there seems no doubt that they must – and if the level of their activities is increased by developments such as THORP (not to mention other factors like the growth of international terrorism), which also seems generally agreed, then public support for these activities seems essential. But the public is not in a position to decide what level of support it will give to what level of activity unless it is informed, in general terms, as to the nature and scope of these activities.

It cannot seriously be doubted that discussion would demonstrate the need for the security services and thereby generate public support for them. Conversely, undue secrecy and lack of accountability encourage a new constituency of critics of the nuclear programme such as Justice and the National Council for Civil Liberties.

Second, security personnel need to know the limits of public acceptability of their activities. A corollary of the failure to debate the role of the security services publicly is that security personnel are offered no guidance as to what the general public is prepared to accept as reasonable behaviour in their part, which would seem to put them in an awkward position. Public debate, even in the most general terms, could serve a useful purpose in this respect.

Third, public accountability discourages abuse of power. The absence of any systematic provision for public accountability inevitably lends itself, at least potentially, to the abuse of power. The revelations at the Watergate hearings, for example, have acted as a salutary reminder in this connection. We are not suggesting that there has been a 'British Watergate', but we do wonder about the extent to which the security services would have been able to impose a cloak of secrecy if there had been one. We note here that the indiscriminate vetting of foreign telegrams by the security services has never been justified to Parliament or the public. We also note that Britain is unique, among Western European countries at least, in not having a statute which sets the framework for the security services' activities, and that an appeal to the European Human Rights Commission in Strasbourg would almost certainly result in a request for Britain to enact such legislation.

Fourth, it can be argued that public scrutiny encourages efficiency. The security services face a difficult and rapidly changing situation in which technological change has transformed their operations, and rendered accurate and high-speed intelligence gathering and analysis arguably the most vital component of our defence system. To maintain their efficiency, it is argued, the security services need the stimulus of external criticism – like any other branch of government. This line of argument is greatly strengthened by what we know of American experience, which suggests that it is all too easy for the security services to devote too much time to data collection and too little to analysis, and that they can be encouraged in this by the ever more potent range of information-gathering devices at their disposal. Again, the fact that surveillance is much more cost-effective than physical security measures can lead to an undue reliance on surveillance – see US Senate (1976) and Rositzke (1975). It is argued that in the UK as in the USA the efficiency of the security services can only be maintained through the discipline of external criticism, at least of their past activities.

What seems essential is that the concerned citizen should be able to reassure himself that the security precautions are reasonable in scope and monitored in practice. This need not necessarily mean that detailed arrangements are disclosed, and the example of BNFL's contract with the Japanese Government may be relevant here. The company refused disclosure on commercial grounds but agreed to allow Raymond Kidwell, QC, for FOE and David Widdicombe for Windscale Appeal to read the documents in the presence of their counsel Lord Silsoe and the Inspector. The objectors' counsel then gave an edited account of the contents to the Inquiry. Probably the key to the acceptability of this strategy to the objectors was that a person known to be sympathetic to their cause was to be allowed to evaluate the position. It is possible that this model could be relevant in connection with the civil liberties issue. For example, the question of how the term 'subversion' is defined operationally by the security services is one that concerns many observers. It may be that assurances from a person *known to be concerned about such issues* could assuage fears without prejudicing security.

It is true that such a person would have to be subjected to security checks but it should be possible to find a person or persons known to be concerned about civil liberties who are also acceptable to security officials. If it is not it would seem that the fears expressed in the RCEP 6th Report are justified now. Public disquiet might be further assuaged by the Parliamentary application of this principle – the inclusion of MPs known to be sympathetic to these arguments on committees such as the Defence and External Affairs Sub-Committee of the Expenditure Committee.

Against this it can be argued that such a procedure is essentially undemocratic, would place the 'trustee' in an invidious position and is unnecessary because the essential *outlines* can readily be discussed in public without compromising security. Advocates of this position claim that, in the context of nuclear developments, the following questions are among those which could and should be publicly discussed:

(*a*) What order of magnitude is envisaged for armed guards for plutonium and other materials, both within the facility and in transit, what types of arms is it intended that they should carry, and what training and instructions will be given to them as to their use?

(*b*) What approximate number of employees may be subjected to detention at the facility, and for how long, if plutonium or other materials are unaccounted for?

(*c*) By what approximate factor will the establishment of this facility increase the number of security checks of all kinds which will need to be carried out, and what opportunity will applicants for employment have to controvert adverse reports which may be generated about them in the course of such checks?

(*d*) What approximate number of employees will need to be submitted to regular or sporadic physical searches at the facility, and how will these be carried out?

(*e*) By what approximate factor will surveillance need to increase, especially in respect of individuals not themselves employed at the facility?

(*f*) What are the existing legal and administrative provisions governing all the preceding questions in the UK, what changes may need to be made in these, and to what extent are or will such provisions be consistent with the UK's obligations under the international instruments relating to human rights law by which it is bound?

Discussion of security measures can threaten their effectiveness. But secrecy can also increase suspicion and discontent, thereby exacerbating the problems facing the security services. This fundamental problem was what the discussion on civil liberties at WPI centred upon. Therefore an acceptable balance has to be found. A formula needs to be found which would permit debate, and legislation is needed creating a framework within which the security services may operate. If this is not done the debate is certain to continue anyway, and to be rehearsed at future nuclear or nuclear-related inquiries with equally unsatisfactory results; and this seems likely to foster suspicion and discontent, hence risking unnecessary opposition to developments in the nuclear power field and, perhaps, an escalation of security problems.

NOTES

1. In a speech at Harlech in August 1978. Unfortunately no record of this speech exists.
2. In the 22 March, House of Commons debate on the WPI Report, Mr Robin Cook, MP remarked that he and two other MPs had visited Peter Shore to urge the calling-in of the application. He then states: 'We are very grateful he took the course that he eventually did. *We know how hard he had to fight to obtain agreement.*' (Our emphasis: Hansard, 22 March 1978, col. 1586.)
3. As it happens, BNFL report they did not know these terms of reference until 19 April 1977 due to some misunderstandings. We do emphasise that the 'timing' issue is not one which caused problems for objectors alone. In interviews, BNFL remarked that they were preparing their case 'in the dark'.
4. Extensive use of the pre-inquiry period was made in the Fox Inquiry into uranium extraction in Australia.
5. Pu 239 has a half-life of 24,400 years.
6. This could well be an exaggeration since it does not allow for foreign contractors sharing the cost of vitrification.
7. It needs to be emphasised, strongly, that the data used here are those supplied by BNFL at WPI. They are *not* in agreement with data from US sources. Sadly, the financial data was supplied close to the end of the WPI so that no reference to UK/USA cost disparities was possible, nor has any been attempted here. This is the subject of further work and hence the figures reported here should not be construed as being, in any sense, definitive.

REFERENCES

International Commis-   *Recommendations of the ICRP* (adopted 17
    sion on Radiological   January 1977), ICRP 26 (Pergamon,
    Protection (ICRP) (1977)   Oxford).

United States, Environ-   *Environmental Radiation Protection Stan-*
mental Protection   *dards for Nuclear Power Operations*
Agency (1977)   (Washington).

Parker (1978)   *The Windscale Inquiry: Report by the Hon.*
                          Mr Justice Parker (HMSO, London,
                          March).

Conroy, C. (1978)   *What Choice Windscale?* (Friends of the
                          Earth   and   Conservation   Society,
                          London).

Mancuso, T., *et al.* (1977) 'Radiation Exposures of Hanford Workers
                          Dying from Cancer and other Causes',
                          Health Physics, vol. 33, no. 369.

Reissland, J., and Dolphin, 'A Review of the Study of Mortality
G. (1978)   Among Radiation Workers at Hanford',
                          Quarterly   Bulletin   of   the   National
                          Radiological Protection Board, no. 23,
                          April.

Sagan, L. (1978)   'The Mancuso Study: A Comment', in
                          Proceedings of INFO 78 Conference, Los
                          Angeles.

Richings, L., *et al.* (1978) *Environmental Radiation Protection Stan-*
                          dards: an Appreciation, NRPB-R 71,
                          April (HMSO, London).

Burn, D. (1978)   *Nuclear Power and the Energy Crisis*
                          (Macmillan, London, 1978).

Pratt, R. (1978)   'The Economics of THORP', M Litt thesis,
                          University of Aberdeen.

Barnaby, F. (1978)   'The   Politics   of   Reprocessing',   *New*
                          Scientist, 6 April.

Johnson, B. (1977)   *Whose Power to Choose?* (International
                          Institute for Environment and Devel-
                          opment, 1977).

Nye, J. (1978)   'Non-Proliferation – a   Long   Term
                          Strategy', *Foreign Affairs*, April.

Bunyan, T. (1976)   *The Political Police in Britain* (Friedman).

Andrew, C. (1977)   'Whitehall,   Washington   and   the
                          Intelligence   Services',   *International*
                          Affairs, July.

US Senate (1976)   *Final Report of the Select Committee to*
                          Study   Government   Operations   with
                          Respect to Intelligence Activities, vol. 1.

Rositzke, H. (1975)   'America's Secret Operations: A Perspective', *Foreign Affairs*, January.

Pearce, D. W. (1979)   'Social Cost Benefit Analysis and Nuclear Futures', *Energy Economics*, vol. 1, no. 2, April.

Starr, C., *et al.* (1976)   'Philosophical Basis for Risk Analysis', *Annual Review of Energy*, vol. 1.

Appendix to Chapter 7
The Calculation of Radiation
Dose Effects[1]

The simple translation of radiation dose-mortality probabilities used in this chapter understates the complexity of the actual calculations. This brief appendix attempts to show how the effects are calculated.

Consider someone exposed to radiation of 5 rems in a given year, call it year 0. The ICRP/NRPB standards would then suggest that five years later, in year 5, the risk of death *in that year* would be $1/2000 \times 1/400$. This is a risk of 1 in 2000 spread over a period of 40 years. This is $1/80,000$ or, using the conventional notation, 1.25×10^{-5}. The risk in year 6 would be the same, and so on to year 45. This risk – i.e. the risk of mortality in any single year – can be compared to the risk of dying *some time* in the 40 years after year 5. This is $(40 \times 1.25 \times 10^{-5})$ or 1 in 2000.

Now suppose a second 'dose' of 5 rems is received in year 2. Then the risk of mortality in year 5 remains 1.25×10^{-5}, but in year 6 (five years after the second dose and 6 years after the first dose) the risk will be (1.25×10^{-5}) from the first dose and *plus* (1.25×10^{-5}) from the *second* dose, or (2.5×10^{-5}). In year 7 the risk of mortality *in that year* will also be 2.5×10^{-5}, and so on. In year 46 however the mortality risk will be (1.25×10^{-5}) because the risk from the *first* dose will have disappeared, while that from the second will remain.

In the same way, then, risks can be added if radiation doses are received *every year*. For 5 rems in each of the years 0, 1, 2, 3, 4, 5, 6, 7, for example, the risk of dying *in* year 12 will be $(8 \times 1.25 \times 10^{-5}) = 10 \times 10^{-5}$ or 1 in 10,000. The risk of dying in year 48, say, would however be $(5 \times 1.25 \times 10^{-5}) = 6.25 \times 10^{-5}$, since the risks from the doses in years 0, 1 and 2 will have disappeared.

It is important therefore to distinguish the risk of mortality *in a given year* (1 in 80,000 for the 5-rem single-dose case) from the risk of mortality over an entire 'lifetime'. The latter will be higher than the former (1 in 2000 for the single-dose case we initially considered).

Where there is *continuous* exposure the risk from *each dose* can be added, subject to allowance being made for any dose 'wearing off' after a given period. Three points are worth mentioning. First, it makes sense to *add* risks in the manner noted above for contexts in which the recipient of radiation can take no 'evasive' action, since the doses are then continuous. None the less, there will be increasing risks from *other* causes as the recipient's age increases. Second, the absolute levels of risk quoted are based on ICRP/NRPB assessments and, as this chapter has noted, these absolute magnitudes are disputed for low-level exposures. Third, some perspective is perhaps gained by considering that even a peak risk of 1 in 2000 (secured by continuous exposure to 5 rems) is small compared to the risk of dying from cancer from non-radiation sources. That risk in a lifetime is 1 in 5.

Finally, if the linear-dose hypothesis is correct, the *peak* risk of dying associated with a continuous exposure at a rate of 100 mrems per year is

$$\frac{1}{50}[40 \times 1.25 \times 10^{-5}] = 1 \text{ in } 100,000.$$

NOTE

1. We are deeply indebted to Dr J. A. Reissland of NRPB and Ron Edwards of Aberdeen University for the basis of this exposition. Neither is responsible for the actual presentation, however.

8 Was the Windscale Inquiry Efficient?

To most people, 'efficiency' conjures up the idea of reaching the right decision at the lowest cost possible. It is no part of the purpose of this report to judge whether Justice Parker reached the 'right' decision about THORP. The idea is to extend the common concept of efficiency to embrace a much wider objective – not just whether the 'truth' was arrived at, but whether the recommendation and ultimate decision by Parliament was based on not just the fullest possible technical, economic and even political information but also on an assessment of the public's view of THORP. To ask the question whether the WPI *in itself* was efficient is not therefore enough. What matters is whether the system of which WPI was part is an efficient means for deciding on issues like reprocessing plant or fast reactor investment or developing a new and major coalmine, or siting a petrochemical plant, building a new motorway, and so on. The Inquiry, in other words, cannot be separated from the entire network of procedures of which it is part. This is to be emphasised for two reasons. First, criticism of the system need not be criticism of an inquiry which is part of that system and it certainly need not be criticism of the persons whose task it was to execute that inquiry and make the recommendations. Second, in suggesting ways in which a system could be improved it will be evident that some of the proposals are for changes in *function*. To say that the WPI did not serve a given function is not therefore to criticise the Inspector or even the whole structure of the Inquiry, any more than it is a criticism of cricket that it fails to follow the rules of football. We noted in the introductory chapter the startling sensitivity of some of the parties to the nuclear debate, a sensitivity that has devolved on even us as persons trying to take an outside look at an existing system to see if it could be improved. Some rebukes have been justified, but most have been based on a form of word-blindness whereby the critic sees what he wants to see and not what is written down.

Without proceeding to an apologia, since none is warranted, it is of more than passing interest that suggestions for *improvement* can be seen by some as suggestions for complete change, 'revolution', or a commitment to changing institutions in such a way that those who oppose official energy policy can use a new institutional system to postpone perpetually actual decisions. It seems worth saying, then, that what is proposed in the next chapter is a set of *modifications* to the planning system – modifications which, while they would delay decisions of some kinds, possibly for a few years and no more, cannot by any stretch of the imagination be construed as a design for the perpetual postponement of decisions about an energy future or any component part of it. However, the over-reaction to the suggestion for change is of interest in itself and is deserving of a few comments.

First, it is evident that some opponents of nuclear power have switched tactics to some degree, away from outright opposition to the argument that there is no urgency and that one should await examinations of alternatives, whether renewable energy sources, or evaluations of fuel cycle alternatives and so on. That switch in tactics has been 'sensed' by the industry so that, in many ways, the future debate is likely to be one of 'urgency' versus 'pause for reassessment'. In itself, then, such a background can be seen to generate responses to further suggestions that the *system of decision-making* (not the investments about which decisions have to be made) should be modified. For modification can involve delay. One could even advance the hypothesis that Mr Justice Parker had just such a concern underlying his recommendation not about the acceptability of THORP but about its *timing*. For to have suggested that THORP should proceed other than immediately would have been to acknowledge the general force of the opposition's case, which had in part been that THORP was not needed, but also that it could safely be postponed until further evidence was mustered to permit a full appraisal. Of course, there were Japanese and other contracts awaiting signature and, once these contracts and reprocessing itself were accepted as 'permissible', then little was to be gained in postponing a decision. The shift of emphasis from opposition to delay is therefore of sociological interest to those who wish to study the phenomenon of opposition, a study that is both much in need and fascinating in its likely outcome. For our purposes, if seeking delay is the tactic of the future then it is right that institutions be designed to ensure that decisions *are* made. To this end, Mr Justice Parker could be said to have served a valuable function in making such a forthright call for immediate development (assuming, that is, that development was the right decision). This view contrasts with many of his critics, but is offered here because the views expressed in this report do attempt to sustain the planning system as the appropriate context for decision-advice, subject to modification. Suitably adjusted, the system can deal with the implications of any given energy policy. It must clearly be judged a failure if it permits perpetual delay. It must, in our view, also be judged a failure if it generates outcomes that are undemocratic or which encourage unjustified delays brought about by other means. These themes are brought up again in Chapter 11.

Undoubtedly, the WPI 'ventilated the issues' as BNFL put it to us. It was a forum for debate and it enabled many to make statements that could have been made elsewhere but to less effect. It also led others into expressing their views, only to have them ignored completely, a point we return to. Nor can there be anything but admiration for a judge and two experts who can sit through 100 days of complex evidence, hear views repeated by different witnesses whilst retaining composure and an evident capacity to see the meaning of much of the evidence presented. One assessor was quite open in declaring that the Inquiry was an 'ordeal'. Anyone trying to assimilate some four million words over such a short period deserves an accolade just for surviving. But if we revert to the earlier point about whether such a *system* is efficient, all these remarks and tributes to inspectors and assessors should surely indicate a hazard warning ahead. The national implications of THORP are far-reaching. The technical issues are complex. The economics alone is beset with uncertainties, while it is difficult to believe that the problem of low-level radiation was really resolved at the inquiry. Is it really the case that future decisions of this kind should rest on the advice of one man and two assessors? In short, *if* Mr Justice Parker's recommendation was right, it seems to us that it was right *despite* the planning context, not *because* of it. If so, then it is indeed a tribute to the Inspector and his assessors, but it is also a very severe warning that any system which relies on one or a few personalities is risking breakdown and inefficiency.

It is for others to judge whether the decision was right or wrong and there will be no shortage of assessments to this effect. Indeed, they have already been made in numerous but brief statements by the main objecting parties, in letters to national newspapers and so on. The submission here is that the inquiry as a *process* was seriously in danger of generating an outcome which could not have rested on a full appraisal of the relevant information. The idea of ascribing national dimensions to a local inquiry context was an error on the part of the Secretary of State for the Environment and, moreover, was an error that could easily have been avoided by using the institution of a Planning Inquiry Commission (with some adjustments to planning law).

Had THORP been assessed at a PIC there would have been a wider array of expert assessors. There would have been the power to instigate detailed studies. Cross-examination could still have taken place. The final report could have been a detailed assessment of the inquiry and the whole process could have avoided the evident haste with which the WPI proceeded. That Mr Justice Parker initiated investigations of his own, no one denies. What is at issue, however, is that there should have been a detailed study of THORP – its economics, its engineering features, its impact on the local environment – *before* the inquiry began. Instead, parts of each of these emerged in an *ad hoc* fashion during the inquiry, frequently with no time for witnesses and even lawyers to assimilate information to the full. This reflects in no way upon the Inspector, but we note that the questions to be raised at the Inquiry had not been fully established *one month* before the inquiry began. We have established that BNFL did not themselves know what evidence to prepare until close to the inquiry start, and that objectors did not know what they were questioning in detail. It is clear that the objectors themselves could not have presented the fullest possible case given the location of the inquiry and given their comparative financial disadvantage. It is a strange and unsatisfactory system that relies on a 'fair hearing' being given to such an important issue in a context in which objectors had to rely, largely, on the gift of one wealthy person – Sir James Goldsmith's gift of £20,000.

In short, if WPI succeeded it succeeded despite the system and by the skin of its teeth. It ran risks that should not be repeated since no one can guarantee that, if the right decision was made, men of the same calibre can be similarly called upon. Even if they could, no one can seriously expect a fair and comprehensive exchange of views when one side is funded and the other is not, and where there is unequal access to information. Above all, the system is inefficient if it fails to ensure that the debate which is taking place *on behalf of* the public fails to inform the public of what the debate is about. Since THORP came forward as a planning application it should have gone to a Planning Inquiry Commission. We have examined the fears that seemed to surround the introduction of such an institution and they must be judged unwarranted. Had the application occurred at a later date, we would have proposed exactly the same thing but with the additional reforms outlined in the next chapter.

A further problem in placing the THORP inquiry in a local public inquiry framework was that the local aspects of the investment were 'dwarfed' by the grander and more national aspects. In any investment decision it becomes important to 'weight' the interests and votes of those affected by the investment. In this case those affected were both the nation and the local people. As argued previously, the local people are the ones more likely to experience any external costs and, if certain conditions are met, are more likely to reap the employment benefits. Their views should therefore carry more weight than those of the general public, person for person. It can be argued that the local consequences were dealt with adequately at WPI. What is at stake, however, is whether in future such a weighted balance could be preserved if further attempts are made to place 'national agenda' issues in the local inquiry framework. To avoid such a risk it seems sensible to make use of the Planning Inquiry Commission with its powers to debate national issues centrally and to hold a local inquiry concerned solely with the local aspects (see for example the local inquiries which preceded the national hearings into the Third London Airport Inquiry).

Other criticisms of the procedures that led up to the WPI have already been noted. The notice given for the inquiry was inadequate and seriously so. The pre-inquiry meeting failed to order the proceedings satisfactorily and failed to define adequately what it was that was going to be debated. Publicity before the inquiry was poor and the documentation disorderly (which was understandable given inadequate notice and the pace at which the hearings proceeded). To suggest that the hearings were 'rushed' even at 100 days' duration may seem odd, but it is clear that time could have been used more efficiently had witnesses not repeated much of the evidence. Media coverage was limited, thus limiting the extent to which the public could have been informed of the proceedings. Coverage was good in *The Times, Financial Times* and the *Guardian* in the sense that there were numerous individual pieces. Not all of them were instructive, occasionally failing to explain the technologies under investigation. Coverage in the 'popular' newspapers was extremely poor despite the fact that one major daily newspaper had been the first to publicise BNFL's proposed Japanese contract and to coin the phrase 'nuclear dustbin'. Magazines such as *New Scientist* and *Nature* took a sound interest and published several major and informative articles on reprocessing and its technology.

The wider informational aspects raise two questions. First, should it be any part of inquiry procedure to encourage publicity? Second, should the resulting Inspector's Report have an informational function? On the first issue, it is clearly adding a substantial burden to an

Inspector's or Commissioner's role to suggest that they should ensure publicity for the proceedings, and probably an improper role. Nor can anyone dictate to newspaper editors what they put in their papers. The solution would seem to lie in a separate process of disseminating information, and this is outlined in the next chapter. Basically, an energy policy commission is proposed which would have this function as part of its role. This would not just be to provide general information for the public but would include the administration of local information centres attached to public inquiries, which would deal with the specific investment in question. There is a clear argument for using language that maximises general understanding without sacrificing accuracy and there is far more scope for this than is generally recognised. Mr Justice Parker certainly exhibited this concern on several occasions, but it would be untrue to say that it is widely reflected in all such inquiries. If other measures can be introduced to increase publicity, all to the good. At WPI the local newspaper reported the proceedings in detail and Radio Carlisle ran a universally acclaimed short summary each day. Arguably this could have been networked nationally, perhaps in the form of a weekly summary. Whether television should be permitted into inquiry procedures is a moot point. The risk, as we see it, is not that participants will 'play to the camera' so much as that the presence of cameras could deter otherwise willing participants from appearing. We note that TV cameras can now be much less obtrusive, not needing special lighting. The whole issue of media coverage is deserving of extended attention.

The second issue is a matter of rules of guidance for inspectors. The WPI Report cannot be judged to be a record of what happened at WPI and this is all the more unfortunate in that many have since judged the *inquiry* on the basis of the *report*. The Report details few of the submissions made and certainly presents a very limited view of some very long and detailed submissions. From the Report, for example, no one would be aware that the energy scenario produced by Gerald Leach (see Chapter 2) was presented in provisional form. It is not once referred to in the Report. Breach (1978) has pointed out that the evidence of the Director of the Conservation Society, John Davoll, also goes un-mentioned. In other cases one is informed that evidence was given but dismissed as being unconvincing or irrelevant (several times without even saying what the evidence was, which raises the issue of why the witness's name was mentioned at all).

Interestingly, the Leach and Davoll submissions had something in common, as had many other pieces of evidence which are not mentioned in the Report or which are dealt with only by implication. Basically, Davoll was questioning the *values* underlying a 'growth' society, which in turn led to the demand for nuclear energy and hence to such plant as THORP. Leach was not necessarily questioning those values, but he was challenging the official view of energy demand forecasts and trying to show that even those levels of demand, or thereabouts, could be met, in his view, by other means. Both in effect either challenged official orthodoxy as it related to values or official orthodoxy as it related to policy. The Report fails, signally, to reflect these challenges. The 'values' debate seemed irrelevant to the Inspector. In keeping with local inquiry precedent, national policy objectives, whether it is a nuclear power programme or the pursuit of economic growth, cannot be challenged. In the suggested modifications of the system we propose that this distinction be made unambiguously clear and that, indeed, the debate about values and about programmes should be removed from the local inquiry framework and be dealt with in national inquiries. This is because what is being assessed *is* national policy. The fact that it is national policy, however, does not mean that it should not be debated at all. Indeed, it is all the more important that it should be. But the proper place for that is in a national inquiry with a wider range of expertise being brought to bear in assessing the arguments. All that is being suggested is that extra-Parliamentary institutions can and should have a role to play in challenging and debating national policy. Our argument throughout has been that this role is justified precisely because such institutions have emerged to meet just this demand. What is not clear is where energy policy is debated in these terms and where the underlying values which differentiate views of energy futures would be heard and expressed. Parliament remains the ultimate decision-maker, but it is also true that the very existence of extra institutions indicates that Parliament cannot be abreast of all that occurs. Indeed, anyone doubting that an information programme should extend to Members of Parliament should consult the record of the debate on 22 March 1978 on the Windscale Report. Some of the speeches, as Breach (1978) notes, revealed a basic ignorance of the rudiments of nuclear power. One wonders how many MPs relied upon the Report alone for the information they did have. Few would have known that the Report and the Inquiry transcripts often read as if they related to different debates.

In this way, then, and for whatever reason, the Report of the WPI failed to serve an informational function to the extent that it could have done. That need not be a criticism of the Inspector – his task was to decide, and to decide on the basis of evidence he regarded as relevant.

But in rejecting evidence, it would have served a valuable informational function had he explained his reasoning.

The style in which a Report is written is very much a matter of taste. It can also mislead, irritate and have effects which, given the sensitivities that exist on the nuclear debate, are counterproductive. Mr Justice Parker set out to give clear answers to clear questions. This is surely admirable. The way in which those clear answers were given, however, has too much of the ring of the judge about them and too little of the informational role required. Of one witness on financial aspects Mr Justice Parker states: 'I need however say no more about his evidence than that I found it unconvincing' (Parker, 1978, para. 9.8). The 'saying no more' is gratuitous since the evidence is nowhere described in the Report. Another distinguished witness clearly impressed Mr Justice Parker. His concern was with caesium discharges to the sea because of the dangers of contamination of fish. His evidence was confined to this issue. None the less, Mr Justice Parker remarks that 'having heard him give evidence, I have no doubt that had he seen any other matter in respect of which he saw any danger to the public he would have raised it. I regard the fact that he raised no other matter as being of considerable significance' (Parker, 1978, para. 10.55). It is, of course, right that an Inspector should judge a witness's credibility. It is distinctly odd that he should construe the *absence* of a statement on evidence he did not give as 'significant' and therefore, by implication, supportive of the case that there were no other worries. In the same vein, one witness on low-level radiation gave evidence which is briefly recorded, only to be commented on by having the Inspector saying that he 'preferred' the evidence of another witness. That witness in turn refers to BNFL's past errors in failing to keep levels of exposure down to 'acceptable levels'. Mr Justice Parker accepts this remark but then declares that since it is the 'intention' of BNFL to keep radiation levels much lower in the future this evidence shows that things will be satisfactory in the future. This may of course be so; but it would be understandable if the reader asked himself whether he could rely on such assurances if the assurances of the past had not been met. Other examples need not be given. They have already been documented in letters to the Secretary of State for the Environment pointing to errors of fact (something that could have been avoided with the reforms suggested in the next chapter) and the occasionally strange juxtaposition of statements made by objectors with the case *for* THORP (see Appendix 2). Others have also drawn attention to curious imbalances in the Report: the paragraph on an episode about radioactive furniture in one Windscale worker's home, for example, or the lengthy admonition of a witness for producing a film allegedly showing the dangers of radiation – a film which appeared to use material from non-nuclear power stations – compared to sometimes extremely brief statements on important submissions. Mysteries and confusions remain and perhaps they are not surprising given such a lengthy set of submissions and words. To take one example, Dr Cochran of the Natural Resource Defence Agency in the USA had given evidence to suggest that the problems of guarding extracted plutonium were serious, so serious that extra reprocessing was not justified. In his evidence he spoke of 'spiking' of fuels before return to foreign countries, a process of irradiating the fuels to make them dangerous to handle by terrorists and/or suitable only for use in the country of origin's reactors. There are, as Parker notes, no provisions for spiking foreign fuels in existing contracts. It is also, again as Parker notes, likely to be a very expensive programme. It is also an unproven technology. So either spiking is not possible, in which case the terrorist risk is positive and of concern, or it is possible and desirable (Parker's own view) but would alter THORP's economics in a manner that no one seems to know anything about.

Lastly, the Report reveals too many 'acts of faith' which others might not share. That no one has ever built a plant of 1250 tonnes p.a. capacity for reprocessing is true. Typically, one would expect a 'scaling-up' to take place. Mr Justice Parker entertained this suggestion as a credible one, only to dismiss it because the relevant BNFL engineer 'showed himself to be a very impressive witness' and had 28 years of experience in designing related plant. The question, as one other witness put it, is not whether one man should sit in judgement on the credibility of the confidence of those responsible for the plant, but whether the public should have the information upon which to base their sharing, or otherwise, of that confidence. After all, for THORP to be publicly acceptable it is necessary to accept the Inspector's *belief* in BNFL's engineering capability for a hitherto untried plant; his *belief* in the success of scaled-up vitrification processes; his *belief* that spiking can, will and should be developed as a technology, and his *belief* that radioactive emission levels would be kept within very limited bounds, having accepted that they had not always been so in the past. Perhaps, suggestively, it could be pointed out that if there is, say, a 0.9 chance of each of these beliefs being correct, the compound probability of them all being correct is 0.66.

If we have dwelt upon some of the disturbing aspects of the *style* of the Report it is solely to draw attention to the fact that style can and does determine impact and to the fact that as the rules of guidance stand, such flexibility of style is inevitable.

Again, it is not our concern to suggest that the style of the Report interfered with the correctness or otherwise of the conclusion. But the way in which such reports are written cannot be regarded independently of the debate in existence. Mr Justice Parker had power to explain and inform. It seems unfortunate that he did not make full use of those powers. For the style is almost censorious in places; and that can do nothing but reinforce an opposition in its view that such reports are preordained to reach a given conclusion. Reports cannot be written to please everyone and certainly should not be written to placate objectors. It can be suggested that they be written in such a way as to avoid the hardening of attitudes that has resulted after WPI.

Far from reducing the distance between objector and proponent, Mr Justice Parker's Report has increased it. No one pretends that consensus can emerge in the nuclear debate. But divergence of opinion can be reduced and inquiries can be used to ensure that the debate takes place within the desirable confines of the planning and political system.

If there is an overriding criticism it is not so much a criticism of the Inquiry itself. It is that the planning system is not amplified to deal with public information and therefore public participation. Some measures which can assist in this are dealt with in the next chapter.

9 The Institutional Future

INTRODUCTION

It will be apparent from the foregoing that some standing commission is required to deal with the issue of energy programme commitments, leaving 'major' energy investment decisions to be debated within a planning framework unless these investments are seen as implying some programme commitment. We have detailed the various bodies that might conceivably take on these roles and have argued that a standing commission is required for the wider energy debate, that a PIC is required for single 'major' investments with national attributes and the LPI (suitably modified) is suited to localised investments. This does not mean that other bodies have no role to play and it is vital to remember that one body, Parliament, must always be ultimately responsible for energy policy and for scrutinising major energy commitments. We would therefore see a legitimate and proper role for a Select Committee, probably the Select Committee on Science and Technology but perhaps a Select Committee on Energy, in scrutinising decisions.

Moreover, whatever body is decided upon, there must be some guarantee that Parliament will debate its advice, recommendations or findings. We have already noted that, but for extensive pressure-group activity, THORP might never have been debated, not least because of the apparent constitutional error in ascribing to a local public inquiry an issue which so many felt had national dimensions (and which the terms of reference refer to opaquely) but which, under planning law, can only be debated if the relevant Secretary of State wishes to risk appeal by objecting parties. It seems fair to say that Parliament has been lax, if not uninterested, in debating energy issues. This view reflects the limited time spent in debating Windscale and the commitment to the AGR programme, and the delays in responding to the RCEP 6th Report together with the quality of the debate on Windscale. As such, it seems essential that any commission concerned with future energy decisions should be 'linked' more carefully to Parliament.

In saying that Parliament is the ultimate and proper body for such decisions it must be recognised that extra-Parliamentary bodies have now emerged which have within their remit matters relating to energy policy. The growth of these bodies would seem to reflect two factors. The first is that energy decision-making requires technological expertise which few MPs possess. The issues at WPI were technologically complex, but it is not clear that they were so difficult that at least the main elements of the debate could not be grasped fairly quickly. The growth of a 'layman's' nuclear energy literature is to be welcomed in this respect. It must be accepted that some issues cannot be made very simple and, indeed, that there is a danger in presenting a complex problem in simple terms. It is not the case, however, that the current 'nuclear debate', in so far as it is taking place, has been accompanied in any significant way by a serious attempt to simplify matters as far as possible without compromising scientific 'truth'. Parliament could be better informed on the issues that were at stake at Windscale and which will be at stake in the debate over CDFR.

The second factor accounting for the growth of extra-Parliamentary bodies may well be concern that, if left to Parliament, issues will be dealt with on a strictly party-political line and subject to all the Parliamentary procedures whereby debate can be cut short and decisions made without adequate time being allotted. Parliamentary timetables are of course crowded and in this respect the various commissions and advisory bodies that have grown up may be seen as 'arms' of Government. In other respects it does not seem too cynical to say that some have served the function of placating pressure groups while at the same time having no real power to influence decision-making. In this way, Government preserves its power to govern, the burden of debate is shifted away from Parliament and the pressure to take action is relieved by having a body which appears to be considering the issues in question. Hence, the formation of commissions may be welcomed by Government and Parliament alike as relieving them of a burden they might otherwise have to bear. Equally, bodies outside Parliament and Government may see the commissions as ensuring that a proper debate takes place and that, if Parliament fails in its duty to the public, then the findings of that commission can be used as an instrument in any pressure-group campaign. The obvious example was the lengthy delay in debating the RCEP 6th Report of 1976 and the response to it in the 1977 White Paper. The fact that the Report existed could be and was used by outside pressure groups to ensure that some debate eventually took place. Government reaction has of course been formalised in the establishment in 1978 of the Standing Commission on Energy and the Environment.

In these circumstances it may seem odd to be suggesting another Standing Commission and at least one PIC. This is to be justified on the reasoning advanced earlier: that a proper debate has not yet taken place; that the public at large has not been involved in what debate there has been; that investments like CDFR are major energy commitments not just in financial terms but in terms of the kind of energy future they may presage; that major energy decisions will have to take place fairly frequently thereafter if a scenario similar to that of the Department of Energy's Reference Case is established as a policy aim; that existing commissions and bodies which might take on these roles are not fully suited to them; and that there is time to establish such institutions. On the negative side, any alternative can be seen as reflecting the view that a decision must, for whatever reason, be hurried through. Any accelerated procedure is unwarranted in the light of its probable consequences. Of course, *if* energy decisions are urgent, there will be costs of delay. Equally, there are costs in rapid decision-making and we see the latter as exceeding the former.

The nature of the suggested Energy Policy Commission is outlined later. First, however, we need to consider existing bodies and their role in energy decision-making. We further need to assess their suitability to the wider role in question. We have already considered the role of Parliamentary Select Committees and this is not considered further.

THE ENERGY COMMISSION

The Energy Commission was formed in 1977 and held its first meeting on 28 November of that year. Its composition is detailed in Appendix 5. The Commission exists to advise the Secretary of State for Energy on a wide range of issues relating to energy policy. Its terms of reference are 'to advise and assist the Secretary of State for Energy on the development of a strategy for the Energy Sector in the United Kingdom and to advise the Secretary of State on such specific aspects of energy policy as he may from time to time refer to them' (Hansard, 28 July 1977, col. 107). Its documents are published, free of charge, and are available from the Department of Energy. Whether the Commission is in itself a valuable body within the sphere of energy policy is arguable, but its existence has been short and it would be early to judge. The sole issue that is of concern here is whether it would constitute the kind of body that should debate energy policy *publicly* and, if not, what role the Commission should play. In our view the Energy Commission cannot act as the focal point for the kind of national debate necessary for energy policy. If it continues to exist, its views could and should be made known to any new commission and, of course, it should continue to advise the Secretary of State directly. The basic reason for its unsuitability for the purpose at hand is that it is heavily dominated by representatives of the energy-supplying industries, the purchasers of nuclear plant, the manufacturers of such plant and the trade unions in the respective energy industries. This is evident from the list in Appendix 5. This does not mean that the deliberations of the Commission are without value: clearly the desires, experience and expertise of the industry are of essential value. It is not however the appropriate body for advising or recommending on the *public* suitability of any commitment to nuclear or non-nuclear futures if that advice or recommendation is to be based on a representative selection of views about such futures. In short, its recommendation would, by and large, be entirely predictable.[1]

THE ROYAL COMMISSION ON ENVIRONMENTAL POLLUTION

The Royal Commission on Environmental Pollution is a Standing Royal Commission that has already produced six reports, under various Chairmen. In terms of energy policy, its 6th Report, produced under the chairmanship of Sir (now Lord) Brian Flowers, is the most celebrated. In wide circles, this report has achieved high commendation and it has been welcomed for striking the first major cautionary note about a commitment to fast reactors. Naturally, in some circles commited to fast reactor futures it has not been so welcomed. Again, however, the sole issue for our purposes is whether this Commission could act as the appropriate body for deliberating on energy policy. Its very terms of reference would seem to preclude it from doing so, although it is true that the 6th Report could be held to have discussed issues outside the confines of pollution hazards – e.g. terrorism and energy policy. Since the chairmanship changes (and has changed since the 6th Report) it is not of course clear that the Commission would report in a similar fashion a second time. We feel, however, that the issues at stake extend beyond (but include) the pollution aspects, of say, CDFR and Belvoir coalfield. They do include the whole issue of energy policy, and whether

CDFR and Belvoir are needed. We do not doubt that the RCEP would wish to provide some input to a new Commission's deliberations. The composition of RCEP (as of 1978) is recorded in Appendix 5.

COMMISSION ON ENERGY AND THE ENVIRONMENT

This was established in 1978 in response to RCEP's 6th Report. Its composition is noted in Appendix 5. It is charged with wide-ranging terms of reference 'to advise on the inter-action between energy policy and the environment'. Appendix 5 also shows that the Commission is linked by some common membership with both the Energy Commission and the Royal Commission on Environmental Pollution. In announcing the Commission, the Secretary of State for the Environment described its wide powers and included the following statement: 'It will also be concerned with planning – though not with specific planning cases – examining the interface between energy policies on land-use planning, and the implications of such policies for the natural world and the environment'.[2]

The CEE began work in 1978. Technically, the CEE *could* take on the role suggested for a national commission which monitors and debates energy policy and programmes. Its concern with the 'environment' may be widely construed, as the term now generally is, to encompass matters outside the natural environment alone. Therefore, if we think of environment as being the 'natural and social environment', the CEE's terms of reference would seem to enable it to discuss the general issues relating to nuclear and other energy futures. Issues such as nuclear proliferation could be held to lie outside the terms of reference, however, although a similar constraint was placed on the Royal Commission on Environmental Pollution but this did not prevent that Commission discussing the threat of sabotage and terrorism in its 6th Report in 1976. As Mr Justice Parker notes, in the WPI Report, there is also likely to be some limitation on the discussion of military aspects simply because this would involve discussing national security. Possibly, therefore, the CEE might be able to range across the broad spectrum of issues which seem to cause concern in the context of energy programmes.

One simple administrative reason for making this suggestion is that it makes use of an *existing*, albeit new, institution. The alternatives are either that an existing Commission be disbanded or that *four* such Commissions, all dealing with energy policy and/or environment, should exist: the EC, the CEE, the RCEP and a new Commission on Energy Policy. It hardly seems feasible, or even desirable, to disband existing Commissions. Hence the scope of one of the Commissions has to be adjusted or a new one introduced. These are the two options which seem to us to be the most desirable. Of the existing Commissions, it seems that the RCEP terms of reference are such that, while it could report again on energy policy, the explicit concern of the CEE with energy makes it *prima facie* more suitable.

A second reason for adjusting the existing CEE rather than producing a new Commission is also purely political. The idea of a new Commission has already been advanced by the Town and Country Planning Association. The only Government reaction is the response of the Secretary of State for Energy on 28 February 1978. There he spoke of not accepting 'the arguments in the [TCPA] statement, because I fear they would weaken Parliamentary control of energy policy'. Perhaps another proposal for a new Commission would not be welcomed.

However, the TCPA statement speaks of a Royal Commission (replacing the Energy Commission) devising an 'energy strategy' in a two-year period *before* any public inquiry is held into CDFR or Belvoir. Energy policy must be the province of the suggested Commission but it does seem improper to make it the Commission's task to devise *an* energy *strategy*, rather than consider alternatives. It may well be this aspect which prompted the Secretary of State to feel that the TCPA Commission, if it came about, would weaken Parliamentary control of energy policy. The functions of the Commission we have in mind are set out later. They are designed to complement Parliamentary control rather than substitute for it.

There are, however, reasons for suggesting that the CEE is not the suitable body for the kind of 'watchdog' role that is suggested for a national commission on energy policy. While it is possible to construe 'environment' in a broad sense, much of the energy policy review process must be given over to the assessment of the state of technology in energy; and it is far from clear that this could legitimately come within the remit of the CEE. The issue of energy-demand forecasting would be another area of concern and this too would seem to lie without the confines of CEE.

A further reason concerns the composition of the CEE. It is, as with other commissions, part-time. It will be evident from scrutiny of the WPI that three experts sitting *full time* for 100 days faced some four million words of submissions and oral debate and deliberations. As is suggested in this report, it remains the case that the fullest debate did not take place. Even if it did, multiplying the numbers of assessors would not have reduced the volume of submissions – it could even have extended the period of the inquiry. Additionally, the WPI was concerned with one, albeit important, investment related to civilian nuclear capability. A commission ranging across the whole spectrum of energy policy would require full-time members representative of a wider spectrum of expertise than is available even to the RCEP and the CEE.

Also, the CEE has already established its first priority – a study of the coal industry to take 18–24 months. This choice was in part based on the feeling that the nuclear industry was already being covered extensively – it needs to be recalled that the CEE was set up in response to the RCEP 6th Report on Nuclear Power and in establishing it the Government's initial idea was that the CEE should review nuclear energy. The choice of coal is also sensible in that coal is likely to form an integral part of any energy scenario. None the less, even if the CEE sees nuclear power as being adequately debated elsewhere, it is not a view that is accepted in this report. This in no way suggests WPI did not perform an extremely valuable function. It is a function that needs to be sustained and extended.

Like other Standing Commissions the CEE can and will carry out investigative studies but it technically has no budget of its own and would have to recommend such studies to the relevant Departments.

Lastly, there is the issue of considering alternative scenarios. We saw that this generated a problem at WPI, with alternative scenarios being listened to but being dismissed effectively because they constituted a procedurally inadmissible challenge to Government policy. This must be an integral part of a new Commission's activities. The CEE's proposal of looking at energy and the environment industry by industry, while immensely valuable in itself, is not entirely consistent with this requirement.

Accordingly, there is a need for a new commission: a National Commission on Energy Policy.

AN ENERGY POLICY COMMISSION

It is evident from the previous survey of existing institutions that there already exist several advisory bodies which have as their function the continuous review and assessment of energy policy, either as a policy in itself or in respect of its implications for the environment. It was argued, however, that none serves the function that needs to be served. What is suggested is that there should be a commission with linkages to Parliament, as the decision-making body, and to the planning system as the main extra-Parliamentary set of institutions that serve a recommendatory role. If these arguments are accepted then we have, admittedly, 'proliferated' commissions and it invites the response that yet another commission is just one more way of consigning issues to a vacuum. It could also be suggested that yet another commission would add to the delays in the system already and that we cannot 'afford' delay. Finally, there is a marked problem of so defining the role of another commission as to inhibit overlap with other commissions.

The justification for a further commission must however rest on the fact that none of the existing commissions can serve the functions we have in mind. When these functions are spelt out in a little more detail this should become clear. Further, while the RCEP, the EC and CEE are all 'standing' commissions, as a new energy policy commission would be, they are staffed by part-time members. Their work is undoubtedly of value and the RCEP Sixth Report has done a great deal to establish a better atmosphere for debate on nuclear power. The same atmosphere now needs to be generated for other energy sources, including alternative energy sources. The charge that commissions inevitably exist to consign recommendations to a policy limbo has some justification in the light of the history of Royal Commissions and the like. But in part this reflects a failure to link those commissions to the political and planning system, and in part it reflects an image of commissions past and not commissions in recent years. Notably, the RCEP Sixth Report is being acted upon in many respects. There is every reason to think that a new vigilance by the public and by Parliament is the order of the day. On the issue of 'delay' it must be accepted that an extra commission would probably delay matters if, and only if, the recommendations of that commission are sufficiently forceful to generate a 'rethink' of a given policy as, we have argued, has been the case with the impact of the RCEP Sixth Report on the idea of a fast breeder reactor *programme*. The question is, then, whether further delay can be afforded. According to the Green Paper on Energy Policy it can be so afforded with respect to a programme of nuclear reactors. The last programme of AGRs is still in the completion stages, while two new ones have been ordered (Heysham B and Torness). The kind of delay envisaged in setting up a new commission is not therefore of a magnitude that would 'harm' a nuclear programme – it is a matter of two or three years, not five or ten. Moreover, a new commission can work very much faster than any of the existing commissions simply because it will, in the modified system we propose, be comprised of full-time paid members. Moreover, in the early stages it may well be that the planning system as it now exists must be left to deal with investments without their 'national' aspects being referred to a new commission. Even here, however, the system has the means of dealing with investments that qualify for the 'national agenda' – the Planning Inquiry Commission exists for that purpose if suitable small changes are made to the Act.

It would seem that the only serious issue relating to the practical desirability of a new commission is its overlap with existing commissions. This overlap does not seem to have been a worry to date – it is far from clear, for example, how the line is drawn between the new CEE and the RCEP given that energy conversion accounts for the major part of pollution. One 'solution' would be to replace the existing Energy Commission with the new commission. As a possibility, this should not be ruled out. None the less, as noted, the Energy Commission has had little time to establish itself and 'make a mark'. Certainly, to date, it is unclear that it is acting other than as a 'rubber stamp' for policy internally agreed and presented to it. But perhaps this is uncharitable and time will tell. Moreover, there is a case for having a commission that is dominated by the industries involved if only to avoid the kinds of mistake involved in not letting the purchasers of energy plant (reactors for example) have as much of a say as perhaps they should have done. The same may be true for trade unionists. Accordingly, it must be accepted that the modified system does involve 'proliferation' of commissions. There is a cost, but the benefits must be judged to outweigh these.

What would such a commission do? Its central role would be to review energy policy in all its aspects – actual programme decisions and R&D programmes. The way in which it would be involved with single investments is dealt with in the next chapter. By and large, single investments would remain the 'province' of the local inquiry, but the new commission would have the capacity to send a representative to an inquiry in which it is judged that a 'national' issue may be at stake. Further, the commission would send a representative to an inquiry in which the commission's own information programme – to be discussed – was queried. Attendance at such inquiries would be in terms of an observational and informational role. In no way would the presence of a Commissioner occasion a challenge to the Inspector's powers.

The new Energy Policy Commission (EPC) would have the function of 'monitoring' local planning applications to see if they are likely to call for an inquiry format that exceeds the local level, e.g. a PIC. This monitoring function need not replace that of the Department of the Environment but it would place it on a systematic basis instead of the current system whereby a planning application may not actually reach an inquiry stage even if deserving of one unless those concerned happen to be aware of its existence. Where the Commission feels an application deserves closer scrutiny for its 'national' dimension, it can then draw the attention of the Department of the Environment or Department of Energy to the existence of this application. The final decision to 'call in' an application would then rest with the relevant Department, as it does now. Alternatively, if it is felt that the Department cannot be the initiator of an investigation and the judge of that investigation's recommendations (as has been accepted in the case of road inquiries), then the Commission can draw its monitoring results to the attention of, say, the Lord Chancellor's office. In time it may even be the case that the Commission would both monitor and advise the relevant Minister and then have the responsibility of appointing inspectors. This development may be undesirable, however, if the Commission also sends observers to the inquiry.

The Commission would also be linked to the planning system in other ways. First, it would have the responsibility of laying down rules of guidance for the funding of objectors. The next section describes the way in which it is suggested the legal status of objectors be changed and argues that they should be funded on a discriminatory basis. It is the EPC that would administer this funding in association with the relevant government department. The Commission would also take on a most important function: the dissemination of information.

It was noted that information is the key to an efficient decision-making procedure, whether it is an MP or the 'man in the street' who needs informing. This information 'programme' has been central to the proposals we suggest and to the criticism of existing institutions. The EPC would have the function of administering the information programme designed to reach the public at large. This is not intended to replace the material supplied by the AEA or the National Coal Board, the Friends of the Earth, or whoever. It is designed to present information in either a 'neutral' fashion or, more likely, in a 'dialectical' fashion with arguments for and against. The Commission, if it has a consensus view, can and should present that view as well. In this way, what both sides of the energy debate call 'education' and 'information' – and which in reality is persuasion – would be supplemented by an independent source. This information could take the form of leaflets and pamphlets, even books. There is no reason either why it should not take the form of broadcasts and television programmes, advertisements or whatever. In the latter respect a modest beginning might be national requests for people to write in with their views and worries.

In such a national programme of information it is essential that the Commission be charged with the wider task that has so far been omitted, largely, from existing energy debates. It should state what the *cost of going without an option* is in terms of forgone energy and national product. It should also state what the *cost of choosing an option* is. It is evident that few people appreciate the 'trade-offs' in the energy debate. To this end, the Commission should consider alternative scenarios, whatever their source, and seek to evaluate them. As noted earlier, the energy debate so far as it has taken place has been conspicuous for its absence of scenario descriptions and their implications – political, social and economic. Had such assessments existed we believe that the debate at WPI might have been better structured and better understood. The local inquiry is not a sensible locale for debates about 'desirable' or 'plausible' futures. But informing people about the nature of that debate can usefully be fulfilled by the EPC through its information programme.

As a further link to the planning system it is also suggested that the EPC be responsible for the local information centres which should exist for local inquiries and planning inquiry commissions (see next section). The Commission itself would have to have a secretariat and this should not be staffed altogether by government department personnel. The reasoning here is that independence from Government must be assured for the Commission to command credibility. Its members must be full-time and also not drawn from government. Access to government officials should be automatic and would arguably be greatly improved if a Freedom of Information Act existed. The Commission would be funded by Government but, like Royal Commissions, would not be other than nominally accountable for the way its funds were spent. It must however honour its functions of reviewing energy policy, instigating studies relevant to that review, and operating an informational programme. The size of membership is debatable. Commissions of more than twenty persons would seem unwieldly and a figure of ten plus a secretariat would make more sense. Co-option of experts for limited periods would be possible. Lastly, no particular composition is sought, but the linkage with Parliament remains. It is suggested therefore that, while such persons could not devote themselves full-time to the Commission, a Member of Parliament, or more than one, should exist as ex-officio members. The aim here would be to admit of the ready translation of the Commission's deliberations to Parliament in that questions might then be asked in the House of the Member(s) so delegated.

To sum up, Figure 9.1 shows the proposed EPC in schematic form along with the kinds of issue that would be debated locally and nationally.

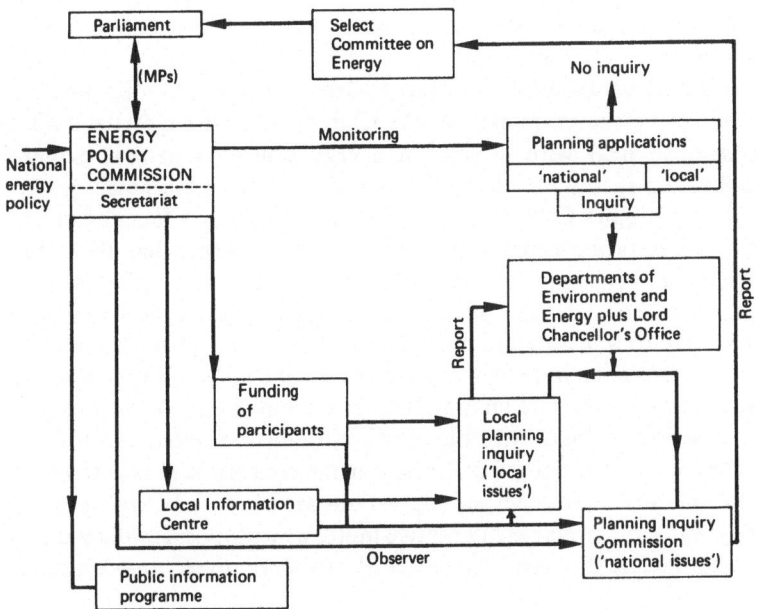

FIG. 9.1 The Energy Policy Commission – issues and links

REFORM OF PLANNING PROCEDURES

(a) Local Inquiries and National Issues
The local planning inquiry remains, and should remain, an integral part

of the process of deliberating on issues which fall into what we have called the 'local agenda'. Increasingly, however, there is the threat, formalised at the WPI, that they will have to bear the burden of deciding on issues which have national attributes. To demand of a local inquiry that it debate and evaluate these national attributes is to ask more of this machinery than it can provide. The appropriate place for national issues is at national level, within Planning Inquiry Commissions, in Parliament and within the non-Parliamentary commissions we have described and recommended. In short, what should have evolved as a procedure for recording the votes, feelings and aspirations of citizens within the local jurisdiction threatens to become an arena for national policy debates. These latter should, indeed must, occur, but the local public inquiry is not the place for them.

It would seem sensible therefore to debate the national dimensions at a 'higher-tier' level, namely in the Commission we suggest should be formed to deal with advice on energy policy. Where a planning application is a multi-site application it can then legitimately go to a Planning Inquiry Commission: again it should not be a function of local inquiries to propose new sites or alternative sites other than the one for which the local inquiry is constituted. This does not reflect a major departure from existing practice, although it does suggest that those local inquiries where alternative sites have been proposed (see the Bacton case cited in note 4 to Chapter 3) have taken on functions that are better served by a Planning Inquiry Commission. Nor is it likely to be in serious conflict with the decision to allow alternative routes for highways to be proposed at LPIs. For the concern here is with energy decisions and alternative site suggestions are likely to be widely spaced geographically, whereas alternative motorway proposals will tend to be not so distant (the road has, after all, to serve an origin–destination function).

PICs need not be unduly lengthy. None has yet occurred, but it is possible that a PIC could be a well-contained, efficient inquiry process lasting less long than some local inquiries. For other investments, such as CDFR, and for multi-site applications for major investments such as petrochemical works, gas terminals, etc., the PIC is most suited and may well be 'elaborate' by comparison with local inquiry procedures. The essential feature remains that *local* inquiries should not be charged with national terms of reference. Adequate debate on the latter requires investigative and research powers which an inspector, with or without assessors, does not have; it requires an array of skills in different disciplines which the inspector may not have and indeed is very unlikely to have. Further, it risks the 'suppression' of the local interest to the wider, sometimes more 'glamorous', national issues. It also forces a dilemma in which the 'need' for the investment can involve challenges to official policy which a local inspector may deem illicit or outside the limits of his capability. We note that the WPI illustrated this feature fully in its determination of the admissibility of evidence which challenged official energy demand forecasts.

Yet, obviously, local decisions aggregate into national decisions. A series of negative decisions on the siting of, say, future nuclear reactors would interfere with Government policy on the growth of nuclear power. There may be no ultimate solution to this problem, if it occurs, as we illustrate more fully in Chapter 11. Partly to ameliorate this problem it is not enough to have official representatives of the involved departments there to state what Government policy is. In addition, it would be helpful if at least one 'observer' from the suggested national Commission be present at inquiries where it is evident that national policy will be questioned. His or her role is not then to present national policy since he or she is independent of it. Nor is it to sit in judgment, although it may be valuable to permit inquiry inspectors to take advantage of the Commissioner's presence by seeking expert advice or the Commission's views on wider issues (note, for example, the use of the Royal Commission on Environmental Pollution's 6th Report at the WPI). The main aim would be for the Commissioner to 'monitor' local inquiries where national problems arise, so as to feed back reactions to the national Commission.

Moreover, the national Commission is charged with an informational role. That information should be made freely available to participants in local inquiries. Again, the presence of a Commissioner in a purely observational role would assist in the event that this information is challenged or queried. Finally, by having Commissioners appear at some (not all) local energy inquiries we have a hitherto unforged link between the local inquiry and a national body. As we noted, the Commission would also be linked to Parliament through membership and hence the full linkage exists from local inquiry up to the final decision-making body. This formation of a linkage may appear unnecessary. It is there because of a view of how future inquiries will develop. That view may be wrong, but even if so the cost of the modifications suggested is slight. If it is right we would expect to see more and more attempts to invoke 'national' dimensions at local inquiries, whereas the aim is to separate the two while retaining an informational flow between the institutions that debate the 'national' and the 'local' agenda. Further as Chapter 11 argues, there is every prospect that future local decisions will, in aggregate, conflict with national policy. There can be no pretence that establishing links of the kind suggested will overcome this problem if it occurs. It may however assist in the event that any series of local decisions has to be overruled centrally for a programme to take place at all.

In short, the aim is to restore to the local inquiry its traditional role of dealing with investments where local public aspects are so dominant that the 'votes' for and against that investment reflect, or are weighted heavily by, local views. The rationale is simple: it is they who reap the external costs if they occur, even if it is the national taxpayer who bears the money cost. Often, though certainly not universally, it is also the local community that reaps the local employment and income benefits. Their votes should therefore count for more.

(b) The Pre-inquiry Period
No local inquiry or PIC can be efficient (in any sense) if sufficient notice is not given to participants. We observed in the WPI case that adequate notice was not given. A minimum of six months' notice should be afforded and no less than three months from the inquiry date at least one pre-inquiry meeting should be held. At that meeting the Rule 6 statement from the relevant Secretary of State should be available. This should not be too much to request since most statements will follow a general format. Where the format is changed it will be because the relevant Secretary of State has so decided, perhaps on advice from the national Energy Commission which, we propose, should also have the role of 'monitoring' local inquiries for their 'national' aspects. Also at the pre-inquiry meeting the party making the planning application should place before the participants and in a local information centre specially established for the purpose all documents known at that time to be relevant to the inquiry. Since the full terms of reference will only be worked out at the pre-inquiry meeting these documents may not all prove to be directly relevant. Within one further month, therefore, the applicant should have prepared a full statement covering the siting of the facility, its engineering characteristics, its cost, its environmental features, and its impact on employment. It matters little what such a document is called, but it will be evident that it is like an environmental impact statement. It is suggested, however, that it should not include any prerequisite for the employment of a particular methodology – e.g. cost-benefit analysis. While cost-benefit is useful for 'ordering' topics for analysis and evaluation, it carries with it the temptation to engage in an

activity which, instead of eliciting the underlying values of the participants, obscures them (Pearce, 1978; Nash *et al.*, 1975). In the presentation of documentation a further requirement is suggested, analogous to that used in civil litigation, namely, that the 'law of discovery' be applied. This would place an onus on the applicant to present all documentation in support of his case *and* that documentation which is in his/her possession and which is prejudicial to the case. This requirement should be made in the absence of a Freedom of Information Act. Where such an Act exists, however, the suggestion would be that it would largely remove the need for the implied suspicion that the requirement devolves on the applicant. By and large, a Freedom of Information Act extended to cover public enterprises would remove many of the unfortunate features of public inquiries which inhibit public participation. Note that the requirement does not involve producing, say, every criticism of nuclear energy to be found in the UKAEA's main library (if they are the applicants!) but only documentation relevant to the site and investment in question.

The pre-inquiry meeting should also be used to 'order' the events of the inquiry by topic. No one can seriously expect a given timetable to be adhered to rigidly: evidence is produced as the inquiry proceeds and new witnesses and evidence may emerge. It is important, however, to make extensive use of the pre-inquiry meetings to eliminate as much overlap in evidence as possible, to ensure that objectors making similar points be asked to meet together to consider whether their evidence may be jointly presented, and so on. It is recognised that this means that parties who wish to speak on more than one topic may be asked to attend several times, but we argue that this disadvantage is outweighed by the elimination of much repetition. This is an area where financial assistance to objectors could help produce a more orderly and perhaps briefer inquiry.

The pre-inquiry period should also be used to establish a local information centre. This may happen 'spontaneously' in that the applicant may offer facilities to assist objectors (as BNFL did at WPI), but local and independent information centres, set up prior to the inquiry *and* for a period *after* the inquiry should be established as a matter of course. The information therein should consist of documentation relevant to the inquiry or, if it is not immediately available, a service for securing it should be provided. Notice that this principle has been accepted, for an experimental period anyway, with road inquiries. The service should be free, however, so as to remove the disparity between applicant and objector in their access to information. Since the kinds of information that would be stocked in such centres and the kinds of advice that it will be necessary to give will eventually tend to be similar for energy installations, one can confidently expect an expertise to emerge within such 'travelling centres'. It is further suggested that to ensure independence, these centres should be under the aegis of the national Energy Commission to be established. Indeed, it is the Commission that should have the power to devolve information to the centres and to the inquiry itself.

(c) Information

The issue of information has generally been dealt with in the discussion of the national Commission. It is that body which would have the responsibility for the information programme designed to reach the wider public through leaflets, pamphlets, advertisements in newspapers and even television programmes. This informational function should be removed from Government or its agencies to ensure that 'fairness' is seen to apply. For this, as we saw, the Commission must itself establish its independence from Government. It would serve the function of an Ombudsman, with the important differences that it would act before and not after events and have more extensive powers.

There is however a feature of informational flows which affects the local inquiry itself. Typically, an Inspector's role is to hear evidence, to ensure that an opportunity is given for cross-examination and then to report on that evidence which, in his view, relates to the advice in hand, usually whether to grant planning permission or not. It is evident from inspection of various local inquiries that considerable flexibility exists in the way in which an Inspector may construe his rules of guidance. The rules of guidance for the setting out of his report are public and those that give guidance on the procedure for inquiries are also now public.[3] As a minimal step, therefore, *all* rules for guidance should be made available at the pre-inquiry meetings and in the local information centre. Otherwise the conduct and functions of inquiries must remain unclear and a source of debate and confusion, one that wasted much time at WPI.

The informational role of the Inspector's report should be formally established in the rules for guidance. It has been argued that the report of the WPI is not a record of the proceedings. It did not have to be, on the interpretation of many. We have already suggested however that a debate of such importance should be conveyed to the public. If an Inspector then wishes to dismiss some of the evidence this is in keeping with the dominant requirement, that he give advice to the Minister on how to proceed. But, whereas Inspectors' reports have not traditionally been mandated to state what actually happened in an inquiry, the suggestion is that such a mandate should exist. This is aimed at extending the purpose of the inquiry to serve an informational function. It need not conflict in any way with an evaluation of the advisory function the Inspector has to perform. The two basic reasons for suggesting that this new function be made mandatory are, first, that a summary of the evidence for and against would inform the Minister and assist him in judging whether the Inspector had effectively selected that evidence which is most relevant.[4] This does not imply mistrust of an Inspector. It does imply that there should be a check on what evidence an Inspector has regarded as relevant and what evidence he has rejected on grounds of error or irrelevance. At the moment there is too much scope for omitting reference to evidence altogether on the grounds of alleged irrelevance in such a way that, since the Minister relies almost totally on the report alone, he has no way of checking whether he has an accurate record of what transpired. Moreover, to assign the task of deciding what is and what is not relevant to one man is to place a substantial and perhaps unfair burden on him over what can be major issues, a burden that has no real parallel elsewhere.

The second reason for mandating that a report be an accurate record is that it serves the direct function of informing both those who attended and those who did not and of showing how the Inspector has reached his decision and by what process. To be clear, many Inspectors' reports meet these requirements. Many do not. It seems reasonable to charge the Inspector with a twofold function of advising on a decision and aiding in the process of informing the public, the Minister and Parliament alike.

A further modification to post-inquiry procedures would assist this process, namely the publication of findings of fact before the report goes to the Minister. This provision is allowed for in Scottish law (though at the moment not widely used) and should be extended to all inquiry procedures. It would also assist in establishing the difference between fact, the Inspector's interpretation of fact and his comment on that interpretation. More than occasionally it is not possible to distinguish which is which in reports.

The crucial point is that an Inspector's report is frequently the only *public* document available to those who have not personally participated in the inquiry. From the participants' standpoint, therefore, they can see how a report treats evidence of which they are aware. For those not familiar with the evidence there is a clear temptation to treat the report as a summary of the evidence and therefore to judge the evidence on the basis of the report. The transition from this inescapable context (unless one goes back to the transcripts and submissions) is to one in which there is a clear bias towards agreeing with the contents of the report. This should be avoided and would be avoided if the Inspector was more clearly charged with an informational function so that his report is a summary of the evidence presented.

(d) Status of Parties
The different 'legal' status of objectors has been described in some detail in Chapter 5. The suggestion is that all objectors, statutory and non-statutory, be afforded the same legal status. This seems a logical extension of some previous rulings on the status of non-statutory objectors and is, we suggest, in keeping with the formal recognition of social changes that have occurred anyway. It seems logically odd to retain a difference in status when public inquiries are almost universally dominated by 'professional' groups of objectors, or experts called in by local residents. To be clear, what the conferment of such status would do is to permit the use of the Courts to challenge an Inspector's ruling on the recommendations he makes but only in respect of errors of law. The reasons why such conferment of equal legal status would not give rise to extended proceedings were rehearsed earlier. Lest there be any misunderstanding, what is *not* being suggested is that the investment decision itself be a matter for High Court judgment. Indeed, much as many objectors would like to avail themselves of the opportunities afforded for indefinite delay to investments that could be brought about by the use of the Courts, as in the USA, it would be our view that the use of the legal process in this way is thoroughly undesirable. The public inquiry exists for the purpose of having the debate. It is none the less desirable that what are currently fairly unchallengeable powers to report on proceedings, in a manner which could even be prejudicial to participants, should be modified to move the balance a little in favour of objectors. Moreover, the removal of differential legal status would merely formalise existing practice, in which no person with a legitimate interest is denied the right to appear. That an Inspector's rules of guidance should none the less enable him to dismiss an 'unreasonable' witness is clear: no purpose, other than a disreputable one, is served by permitting disruption of proceedings.

The principle of equal legal status would extend to a principle of equal initial credibility. As we saw in the discussion of WPI a confusion emerged when alternative energy forecasts and futures were being discussed. The Inspector's recourse was to acceptance of official forecasts on the grounds that objectors had no responsibility for their own forecasts, thus giving credibility to 'official' forecasts and less credibility to 'unofficial' forecasts. In the event, this unfortunate implication could have been avoided by not hearing evidence on forecasting, for if it constituted a challenge to Government policy then it should have been removed from the local inquiry context. (As it happens it is unclear if it did constitute such a challenge since the forecasts that exist in official circles are contained in a Consultative Document, not a White Paper.)

To further extend equal status we have suggested funding of objectors, although clearly not on a non-discriminatory basis. The rationale here needs restating only briefly. The very fact of holding an inquiry implies that the Minister seeks advice. That advice is the outcome of a process of exchanges of opposing views, evaluated through cross-examination. It is totally unclear how it can be argued that there can be a fair exchange of views when one side, if it is a public agency, is using public money to present its case and the other relies on voluntary contributions and even personal savings (as was the case at WPI). It would seem a sensible role for a national Commission that it should, in monitoring such inquiries, also act as arbiter as to who is and who is not to be funded.

(e) The Inquiry

Lastly we can consider the inquiry itself. The use of the pre-inquiry meeting, at a much greater interval from the inquiry than is usually the case, should permit a more orderly presentation of material by topic than is often the case. No doubt most inquiries proceed at a sensible pace, but where one threatens to last a long time there is a case for 'rest periods' for participants and Inspector to reassess their own evidence and to assimilate new evidence. The main issue concerns the use of an adversarial and 'legal' context. Neither is wholly satisfactory in terms of public participation simply because the very 'courtroom' atmosphere is sufficient to deter the least articulate and the least professional, when, arguably, it is they who should most be heard in the local context. This is especially true of local inquiries. Appendix 1 shows clearly that, with very few exceptions, witnesses giving evidence at WPI were 'professionals'. However, an adversarial purpose is valuable in extracting information where there is any doubt that full information is being brought forth. Introduction of the 'law of discovery' would assist. The greatest assistance, however, might come from a Freedom of

Information Act which could arguably dispense with much of the need for legal representation and the unnecessary sharpness of exchanges which characterise so many public inquiries. Unless either of these changes can be introduced, however, there seems little option but to maintain the adversarial process with lawyers being used where any participant feels one is necessary. Admittedly, once one participant secures legal representation others may feel disadvantaged in not having it. Thus to accept adversarial proceedings may well be to accept legal domination of the process as well.

Again, the presence of lawyers and the courtroom atmosphere can be conducive to securing information, and we noted that this could be important in contexts where mutual suspicion exists and is likely to remain, as in debates over nuclear power. It can only be recorded however that just as an inquiry runs the risk of being a debate between 'elites' so it also runs the risk of being ill served by the legal profession, whose training in the articulation of cross-examination may be enviable to those who see virtue or satire in it, but can be inimical to the pursuit of truth. It is a sad reflection on many inquiries that evidence and opinion are omitted simply because a witness will not come forward into a context where the very language and formality are alien to his or her own mode of expression. The introduction of a Freedom of Information Act could do much to eliminate the need for such tactics to encourage public participation.

One further essential issue on the reform of the planning procedure relates to the need for planning advice to be linked to Parliament. At the moment, an Inspector's report is submitted to the relevant Secretary of State, usually the Secretary of State for the Environment, but increasingly one would expect the Secretary of State for Energy to be involved, or both together. It was seen that the Windscale Report was submitted to the Secretary of State for the Environment, who then faced a demand by MPs for the Report to be debated *before* any decision was made on the planning application. If, however, new evidence appeared during the course of any such debate, objecting or 'defending' parties to the original inquiry could legitimately request a reopening of the inquiry. To avoid this, as we saw, the procedure in the WPI case was for planning permission to be *refused*, for the Report then to be debated in Parliament, for the Secretary of State to *then* introduce a Special Development Order for planning permission and to achieve the award of permission after a minor debate and a vote.

Now, *any* report from a local inquiry will face exactly the same procedure if Parliament insists on a debate of the findings before permission is given (or refused). It is possible that a given Secretary of State would risk the 'new evidence' argument – indeed there is much to be said for suspecting that Mr Shore was over-reacting in seeking the route he did on the Parker report. None the less, if the fear is there and if the issue is important enough, the farce of initially denying permission when what is really meant is that permission should be given will be repeated.

It has already been noted (see Chapter 5) that there is some doubt as to what the procedure would be for the findings of a Planning Inquiry Commission. Essentially, such a Commission would produce a Stage One report on the principle of the investment in question. The legislation is so vague that no one appears to be sure exactly what would happen. Would a Stage One report be debatable by Parliament *before* the local inquiry stage? It would seem that it *could* be, but that the 'spirit' of the planning legislation is that it should not be and that the entire inquiry procedure would be expected to be complete before the issue went to the Secretary of State for reporting to Parliament. In short, the oddity that arose with the Windscale Report would be repeated with the findings of a PIC Stage One and, of course, with the findings of the subsequent local inquiries.

Some escape from this situation is desirable, not just because of the reflection on planning and Parliamentary procedures but, more important, to link planning procedures firmly to Parliament. To achieve this it is suggested that any report relating to 'national' energy matters be routinely given to a Select Committee of Parliament and *not* to the Minister. If the Select Committee judges that no debate is required it will make no call for such a debate, merely recording the receipt of the application in its general reports. If it feels there is a need for the debate then it can, in the process of passing the report and its own recommendation to the Minister, call for this to take place. In this way, the planning process would be linked formally to Parliament in so far as that process related to items judged to be on the 'national agenda'.

To be clear, what is not being suggested is that all Inspectors' reports, from whatever source, be routed via the Select Committee. The idea would be that in a 'refurbished' planning system, national issues would, in their planning context, go to PICs and not to local inquiries. If, however, a local inquiry did take place and was felt by the proposed Energy Policy Commission to warrant further consideration, special application could be made to route that report via the Select Committee. Further, it is not and would not be within the powers of the Select Committee to pronounce judgment on the report and its recommendations. That power must reside formally and firmly in the hands of the relevant Secretary of State. The aim is to ensure a Parliamentary monitoring process whereby those energy decisions with national attributes have the *capability* for Parliamentary discussion, at least within the Select Committee, should the Select Committee so judge. The powers of the Secretary of State would not be reduced.

Such suggestions have been made before and we entertain no great optimism that they will prove politically any more acceptable now than hitherto. Moreover, if the proposal is to work it seems sensible that the Select Committee receiving such reports should be a Select Committee on Energy, not the Select Committee on Science and Technology. Indeed, suitably extrapolated, the idea is generalisable across all investments with 'national' dimensions provided there is a counterpart Select Committee in existence. Such a revised structure has already been proposed (Select Committee on Procedure, 1978). It would seem to us to be a sensible modification of the system and one that would considerably serve the interests of open government.

Throughout this chapter we have spoken of planning procedures – i.e. institutional proceedings that come into play (or may do so) if an actual planning application is made. In a recent publication it has been suggested that some investments with 'national' characteristics be debated *outside* the planning context before they go to an inquiry of whatever form (House of Commons, *Select Committee on Science and Technology, 1978*). The particular investment in question in the document is CDFR, but clearly the idea of 'prior' inquiries that occur before a planning application have been made could be extended. Such a proposal has a distinct advantage in securing a further link between the planning and political process, a matter that has concerned us throughout. Note that we have already suggested that 'national' inquiry reports be submitted to a Select Committee rather than directly to the Secretary of State. The idea of *prior* discussion adds a further dimension to this. It could, of course, duplicate, proceedings since the Select Committee's deliberations are not necessarily inconsistent with a subsequent planning inquiry. Indeed, such an inquiry has to take place if the relevant local authority objects, if sufficient objectors object or if the applicant objects to refusal of planning permission. The Select Committee see no objections to such duplication and indeed feel it would test public opinion.

The arguments against such a prior inquiry by the Select Committee are, however, ones that give occasion for concern. It would, for example, be open to the Secretary of State for the Environment to give planning permission through a Special Development Order (as was finally the case with THORP) and such an Order requires no planning inquiry. It would almost certainly be the subject of debate in Parliament, but that is all. Of course, an Order of this kind could be introduced without any prior debate within the Select Committee having taken place, but the fear must be that a Secretary of State could use the prior investigation as evidence that an inquiry of some sort had already taken place and that a further planning inquiry was not therefore justified. Arguably, such a tactic would meet with serious political opposition. Arguably, it could describe a very realistic possibility. It remains our view, then, that the Select Committee (preferably a Select Committee on Energy) should have its role as defined in the preceding sections and this role should not extend to the holding of prior inquiries into matters which belong in the context of a Planning Inquiry Commission.

NOTES

1. The Town and Country Planning Association recommendation (TCPA 1978) is that the Energy Commission be *replaced* by a Standing Royal Commission on Energy. While very conscious of the 'proliferation' of Commissions we do not consider that the Energy Commission has had time to establish whether it is serving a valuable function or not and we therefore assume its continued existence. For a similar criticism of the Energy Commission's composition see *Third Report of the Select Committee on Science and Technology*, Vol. 1, p. xlvii.
2. Hansard, 2 March 1978, cols 330–1.
3. We are deeply indebted to the Department of the Environment for providing the guidance rules 'B1' which, to our knowledge, have not been made widely available before.
4. In the case of WPI it seems clear that the Secretary of State, Peter Shore, read far more than the Inspector's Report, as his masterly grasp of the issues demonstrated in the House of Commons Debate on 22 March 1978.

REFERENCES

Pearce, D. W. (1978) 'Social Cost-Benefit Analysis and Nuclea Futures', *Energy Economics*, Vol. 1. No. 2 April.

Nash, C. *et al.* (1975) 'An Evaluation of Cost-Benefit Analysis Criteria', *Scottish Journal of Political Economy*, June.

Select Committee on   *First Report from the Select Committee on*
Procedure (1978)   *Procedure, Session 1977–78: Report and Minutes of Proceedings* (HMSO, London).

10 Whither Nuclear Opposition?

There will be more, not less, debate about nuclear power and other energy sources in the UK in the future. Any analysis of the efficiency of decision-making in the energy field must allow for this factor, as must any prescriptions for change. It must be seriously questioned whether the framework of planning law as it now stands in practice is best suited to dealing with what will be a growing pressure to oppose nuclear power in itself, or at least to exact from the industry standards of safety and environmental quality that exceed those now in existence. The suggestions made in this report are designed to amend the system by which planning decisions are made in order to prepare for such a future.

The prediction that opposition will grow rests on several observations. The first, and most self-evident, is that, unless there is some unforeseen and radical change in energy policy, the nuclear component of energy policy will increase. Its rate of increase, after a number of years of stagnation and lack of orders, must be considerable for it to achieve the supply of electricity argued to be necessary to sustain economic growth rates of the order of 3 per cent p.a. to the year 2000. The forecasts may be revised downwards, renewable energy technologies may achieve sudden breakthroughs, but major changes seem unlikely. Since the object of opposition will increase its rate of growth, so will the opposition. Second, and deriving from the first observation, no small part of the opposition emerges from the feeling that technologies that are ill-understood are being introduced at too fast a rate for 'societal control' to be exercised. Very simply, there is a desire to slow down or pause in order to gather more information and to reassess situations. Certainly WPI illustrated this concern, since the debate there was over a technology that is to be introduced on a scale orders of magnitude greater than any existing plant of which there is direct UK experience. Ancillary technologies such as vitrification also exist on a small scale but not on the scale necessary to 'service' waste from THORP, and so on. Essentially, what would once have been introduced as experimental

plant and upgraded in size in discreet intervals is now being contemplated in one 'large jump'. There is an unnerving element in this kind
of progression and it is felt more widely than in the professional
opposition groups. Indeed, some members of the nuclear industry have
expressed such doubts.

Third, in the UK the nuclear industry has until recently faced little
opposition. The emergence of the environmental movement, for
whatever reasons, has now changed that. There is an element of 'culture
shock' in the reaction of the industry to the opposition – much of which,
no one can deny, is based on misinformation and illogical thinking. In
some ways, however, there has been over-reaction precisely because of
this new situation and that over-reaction has itself been characterised by
outrageous statements that almost demand a faith in technology, rather
than seeking to persuade through careful argument, open disclosure of
information and clear expression of risks. But there has been a process
of increased disclosure, sensible debates are taking place and the general
public is *better* informed even if it is a very long way from being
adequately informed. As the construction of reactors on existing and
'green' sites increases, therefore, we may expect a larger part of the
population to avail themselves of the improved informational system
and a larger part to voice opinions and request assurances. Indeed, this
tendency is already clearly in evidence.

Fourth, if nuclear opposition is part of the environmental movement,
and we have warned against making a one-to-one identity between the
two, then, whatever one's hypothesis about the reasons for the
emergence of such a movement there is no evidence that those reasons
will be absent in the coming three decades and perhaps beyond that.

The question that is interesting is how this opposition will present
itself in the future. For, depending on the direction it takes, it has
implications for decision-making procedures. There are perhaps two
observations or predictions that are pertinent here.

To date, the 'professional' opposition in the UK has been 'well-
behaved', using the weapons of argument and debate rather than
attempting occupations of sites or the use of violence, as on the
Continent. The use of the law courts is denied them, in contrast to the
USA, and this is why planning institutions and the direct lobbying of
Parliament and Ministers has afforded a proper and sensible outlet for
the expressions of doubt, whatever one's views of the nature of the
arguments used. It is that framework that needs to be preserved in order
to avoid the kinds of impasse reached in the USA and, more important,
the kinds of demonstration seen on the Continent. But to preserve it

means that it must be seen to present an 'efficient' outlet for the expression of criticism: it must be possible for the framework to contemplate a set of results that are counter to professed Government policy. That, in essence, is surely what even the worst democracy must achieve to merit even the contemplated use of such a title. If nothing else, the message we have tried to get across is exactly that. The institutional framework for securing energy decisions can, and does, vary between countries. In the UK the combination of planning institutions, democratic national bodies and extra-Parliamentary investigative and advisory units represents a potentially powerful means for making sensible and responsive decisions. If, however, their decisions are predictable regardless of the nature of the arguments put to them then the system will have 'legitimised' other tactics, wittingly or unwittingly.

Those other tactics could be the use of the Courts to challenge, let us say, a report such as that on the Windscale Public Inquiry. This prospect was discussed earlier. The chances that a judicial system will itself be critical of some other judiciary or of a quasi-judicial process is remote. Since the professional opposition knows this, it seems unlikely that there will be any widespread recourse to the Courts in the future energy debate. The lobbying of MPs will continue, as, probably, will participation in public inquiries or Commission hearings. But even if these fail to honour the requirements of 'potential responsiveness' – i.e. if they cannot exhibit the capability of entertaining arguments that question and reject official policy – then it is easy to see that resort to civil disobedience will emerge as a singular weapon of the opposition. To be clear, even if it occurred it would not be supported by many who now oppose nuclear power or monolithic energy institutions. Nor should an attempt to explain how civil disobedience can come about be regarded as a justification of it. But within the minds of those who take such decisions their acts could be seen as legitimate because the planning/democratic system will not have demonstrated itself, to them, as possessing the capability of entertaining alternative views. There is therefore a process of internal justification that should at least be heeded. There is not the slightest doubt that some who would resort to 'occupations', disruption of supplies, delaying tactics and so on will do so without any such process of internal self-questioning. But there are others who will, and one of the objectives of the decision-making process should be to ensure that the outlets for expressing concern on the basis of 'fair hearing' exist in as efficient a manner as possible. If, at the end of the day, there are those who will resort to anything to get *their* message across and who will never accept a majority view, then the system must take care that they are identified and separated for what they are. One of the questions is whether the system as it now stands could serve that function, whether it could not in fact be said to be still slightly tinged with the attribute of self-fulfilling policy discussions. And if civil disobedience is the order of the day for future conflict, then, as we suggested in an earlier paper (Pearce *et al.*, 1978a) there is a very thin line between such campaigns and the risk of violence, whatever the objectives of those who might call for such campaigns. Those who argue that modifications to the planning/democratic system are 'idle pastimes' or prescriptions for never-ending delay in making decisions should then reflect on the impact of such a scenario on an energy programme. The potential for delaying the construction of nuclear reactors, for inhibiting a major mining investment, even for simple interference with the work of officials in the energy industries is such that far worse delays could be experienced than those necessarily implied by a modified decision procedure, especially if the latter modifications contain a 'back stop' procedure whereby the decision must ultimately be made one way or the other.

Now, if this analysis is correct we require some evidence that the professional opposition will, or may, resort to civil disobedience. Unquestionably, the aftermath of the Windscale Public Inquiry, but more so the publication of the Report in March 1978, saw an internal debate within the opposition over future tactics. Anyone speaking to members of the various parties at WPI or to other critics of official policy will know that this has been the case. The existence of this debate was reported in our earlier work (Pearce *et al.*, 1978b). Since then, there have been calls for direct action and the boycotting of future inquiries, especially that promised for CDFR (*New Ecologist*, 1978: *Undercurrents*, 1978). Moreover, other observers of the nuclear debate have reached the same conclusions. In his assessment of the Inquiry, Breach (1978) reports:

> But what they [the nuclear industry] were seeing [at WPI] was simply the first phase of a conflict that now seems certain to embroil them and their countries for a very long time to come. More importantly and ominously, it is a conflict in which there can be no winner on present reckoning. The main protagonists can achieve their ends only at the expense of peace or freedom – or both.

Cornford (1978), in making a plea for open government and an Official

Information Act, concludes his paper with the following remarks:

> Finally, it is important for the quality of decisions that the internal
> debate in Whitehall should be opened up: Whitehall has no monopoly
> of wisdom and in some areas there is evidence of ignorance and
> insularity. Government proposals need to be submitted to the test of
> expert opinion in public debate, especially in those spheres of
> technical and scientific policy which involve irreversible decisions
> with momentous long-term consequences. It is for this reason that I
> do not think that changing the rules, and abandoning the assump-
> tions of private government is a matter for leisurely debate. A
> Concorde or two may not matter much, nor a gothic tax system. But
> spectacular mistakes about nuclear energy for instance had better be
> made with full public awareness of the reasons and the widest and
> most open debate. If that is not granted, then opposition will be
> forced into irrational channels and existing forms of protest such as
> the disruption of planning enquiries will acquire an aura of antique
> gentility.

A scenario of disobedience campaigns and even violence is not therefore
one based on idle observation. As we noted in our earlier work (Pearce
et al., 1978b) we would sincerely hope that such a picture of the future is
hopelessly in error. It would be all the more unfortunate, however, if it
came about all for the price of non-revolutionary changes to the
planning system and its rules of procedure, and for the price of a
Freedom of Information Act which is revolutionary in its implications
for the end of what Cornford calls 'private government' but which is not
revolutionary within the context of what a democracy should look like.

Yet there is another and much neglected scenario which has every
prospect of occurring and which spells a greater threat to the interests of
energy industries bent on expansion. Moreover, it is a totally legitimate
threat since it occurs within the planning system. Quite simply, any
proposal to construct a reactor, sink a mine or site an LNG plant must
secure planning permission from the relevant local authority. In the era
of 'quiet progress' with nuclear power, planning applications and their
award was a fairly automatic procedure. Nor has there been much fuss
over coalmining, not least because new mines have been sunk in existing
mining areas so that the existing 'culture' accepted mines as part of the
community's way of life, and because employment of a specialised kind
was being preserved. But nuclear power is hardly embraced as part of
any community's culture, the exception being perhaps in the very north

of Scotland where the people have lived with experimental fast reactors for a long time. Moreover, their employment is heavily dependent on those industries. And the Vale of Belvoir depicts an extension of a mining area and one which impacts on hitherto 'virgin' land. The eventual disposal of nuclear waste in vitrified or other forms requires suitable geological conditions, best found in parts of the country that again have no immediate reason to embrace nuclear power as part of their local culture, even if they can acknowledge its output, electricity, as being part of their very way of life.

Now, while these are early days, there is every prospect that single communities will increasingly reject planning permission for new sites for power stations or for waste disposal. A micro-political decision of this kind need never be related to the macro-political 'need' because it is always possible to say that the installation or facility should go elsewhere. Self-evidently the process is not one that can be aggregated without having implications of the severest kind for national policy. If national policy is mandated and it includes the construction of x reactors then, self-evidently, those reactors must be sited somewhere.

Yet there is evidence that a process of local rejection is going on. Notably, permission was refused for AEA's test drilling of boreholes in Usway Forest, Northumberland. These boreholes are designed to test the geological structure for the disposal of nuclear waste in perhaps two decades' time. Kyle and Carrick District Council in Galloway, Scotland, called a public meeting on 30 August 1978 before adjourning to consider the AEA's request to drill there. They subsequently rejected the application. Similar reactions have greeted AEA's requests in a number of places.

In Gwynedd, North Wales, an interesting experiment was carried out by local planning officers. A document stating the pros and cons of nuclear stations was prepared and over 800 copies were circulated to local councils, community councils, MPs and other groups. Of those responding with definite views one way or the other, 37 organisations plus one MP were *against* further power stations in the Gwynedd area. Five organisations voted in favour. The planning officer for Gwynedd then prepared a second document detailing the responses he received and considering each possible site in the Gwynedd area in terms of the impact of a power station on the environment, employment situation and so on. Each possible site is carefully linked to the structure plan requirements set out in the county's planning documents. The conclusion was forthright:

It is considered that the development of any site in Gwynedd for a further power station would be contrary to the County Council's policies as contained in the Gwynedd Structure Plans. In view of the comments received on the first Report, and in the absence of any clear overall benefits which outweigh the disbenefits of power station developments, there is no justification for considering amendments to those policies. (Gwynedd County Council, 1978)

For our purposes several things are noteworthy. First, the Gwynedd experiment was an experiment in information provision – the first document states fairly clearly what the costs and benefits of power stations are, although in terms of radiation risks, etc., this type of document would require supplementing by the kind of information programme we have suggested. Second, it was an experiment in public participation by taking the first document down to the level of local and community councils. Great stress is laid on this aspect in the documents. Third, it is an exercise in planning and in 'responsiveness' as we have suggested it. The views of the small local units were taken directly into account in formulating the conclusion quoted above. Fourth, the information and consultation approach used a 'for and against' presentation as opposed to a single-sided approach. Thus, while similar experiments are being repeated in other councils in which planning officers are feeding information to councillors on nuclear power, those that we have seen could hardly be called neutral in their presentation of facts. The Gwynedd experiment is much to be preferred. Fifth, the Gwynedd experiment did not involve consultation with the nuclear opposition in its professional guise. It involved consultation with local community organisations and district councils.

All this said, it will be evident what the problem is. The possibility that conclusions of this kind could be reached by a significant number of local authorities is a real one. While it may be possible to show how further nuclear establishments may be incompatible with local structure plans, it requires only a limited number of such conclusions to be reached for the national policy for nuclear power to be seriously threatened. The spectre of Ministers overruling a number of local decisions could therefore be a real one. Yet it is difficult to see how this could be done on any significant scale. If the results of the Gwynedd experiment are repeated elsewhere, the programme implicit in the energy policy document produced by the Department of Energy in 1978 could easily be more difficult to realise than has hitherto been recognised.

REFERENCES

Pearce, D. W., Beuret, G., and Edwards, L.(1978a)
'Opposition to Civilian Nuclear Power', in *Proceedings of 1978 Conference of the Uranium Institute*, (Mining Books, London).

Pearce, D. W., Beuret, G., and Edwards, L. (1978b)
Windscale Assessment and Review Project, Interim Report (Social Science Research Council, London).

New Ecologist (1978)
Reprocessing the Truth (*New Ecologist*, Wadebridge).

Undercurrents (1978)
Undercurrents, Issue 27.

Breach, I. (1978)
Windscale Fallout (Penguin, Harmondsworth).

Cornford, J. (1978)
Official Secrets – A Right to Know (The Outer Circle Policy Unit, London).

Gwynedd County Council (1978)
The Impact of a Power Station on Gwynedd, Vols 1 and 2 (Gwynedd County Council, Caernarvon).

11 Postscript

On 13 September 1978 Peter Shore, the then Secretary of State for the Environment, indicated that he was, in certain circumstances, prepared to change the planning procedures applicable to major investments, citing coal and nuclear in particular. The full text of his speech is printed below.

THE SECRETARY OF STATE'S SPEECH

Speaking in Manchester, Mr Shore said:
'I have during the past year given much thought to our system of planning inquiries – particularly as it operates on major and complex issues of which last year's Windscale Inquiry was an outstanding example. But in the period ahead there will be other major planning inquiries, and I want to indicate to you my present thinking about how they can be best handled.

'Perhaps it will help to clarify thinking if I remind you of how in the post-war years we have tended to approach the major important planning cases: those which are sufficiently controversial to come to Ministers for decision. Something like 5000 inquiries are held every year, and of these perhaps a few hundred are highly significant to the locality. But only say two or three a year interest, concern and affect the well-being of us all. It says much, I think, for our planning inquiry system and its procedures, that for the 30 years it has so far existed it has on the whole managed to deal with the whole range and variety of cases in an acceptable and satisfactory manner.

'Let me remind you of the three main principles on which our public inquiry system has rested. First, that it is for Government and Parliament to determine national policies against which particular proposals are considered at inquiry. These policies – except traditionally in minerals cases – have usually settled such questions as the basic need in national or regional terms for the type of development in question.

'Secondly, against the background of declared national policy there should be, when a planning application has been refused and an appeal has been made, or when a call-in has taken place, a full, scrupulous, impartial and structured inquiry conducted by an Inspector to consider whether there were sufficient reasons for a particular proposal on a particular site, in all the circumstances, to be allowed to proceed or to be turned down.

'Thirdly, in the light of the inquiry and the Inspector's report it is for the Secretary of State under powers specifically granted to him by Parliament in the Town and Country Planning Act to make a decision.

'As I say, in the great majority of important planning cases that come to Ministers, these principles have proved to result — and still do — in decisions that are effective, fair and accepted. Of course there have always been criticisms and this is inevitable because one party is bound to be disappointed and because the procedure, involving as it does a careful and impartial hearing of all the evidence, irritates those who want quick decisions and offends those who are on the losing side who believe that, if only more and more exhaustive studies could be made, their point of view might have prevailed. These criticisms are, I believe, unavoidable but they do not detract from the general utility and value of the system.

'So much for the established features of the system. But in recent times some critical questions have begun to be asked that previously were seldom if ever raised. I will instance three in particular. First, critics have sometimes questioned whether the need for a development has in fact been properly established. And they have claimed that Parliamentary discussion or Ministerial consideration of the first question of the need has often not been sufficiently searching and thorough.

'Secondly, the critics have claimed that certain major proposed developments have implications and repercussions going far beyond the direct impact of the project itself, that these wider effects are not sufficiently considered and that, if they were, the balance between, say national economic considerations and the effect upon the environment and quality of living, could turn out to be very different from that which the developers claim. Consequently critics have argued for a more thorough and disciplined assessment of the total implications of large-scale developments.

'Thirdly, not only the critics but all those engaged in the matter of public scrutiny recognise that there are some development proposals — and I am referring here to major nuclear innovations — that are in a special category of importance and difficulty, not just because they involve technological judgement of great complexity but still more because they can affect our whole way of life and because they involve issues of utmost importance to the safety and health of future generations.

'These are serious concerns and all of them question the traditional approach to public inquiries that I outlined at the beginning of my speech. In short, we must ask the question: do we in fact sufficiently establish and define need in certain fields, particularly the energy field, when in the nature of things it is difficult to establish a settled and continuing national policy background before a particular proposal is examined at a public inquiry? Can we in fact take, at a given point in time, sufficient account of all the wide implications of major new ventures as they evolve; and are the techniques that we can employ sufficient to help us with them? Is it right to leave to the Secretary of State, in the vital field of major nuclear innovations which can affect all our future the sole decision – or should Parliament be directly involved?

'Some of the considerations I have been outlining will arise on at least two major energy development proposals which will require Ministerial decisions – the National Coal Board's applications for planning permission for the development of a major new coalfield at Belvoir in North East Leicestershire which have recently been submitted: and the proposal – if and when it comes forward – for a fast breeder nuclear reactor, the CFR 1. May I give you my ideas about the examination of these two very different issues in turn?

'The NCB's applications raise issues of considerable national importance relating to the need for the development of this coalfield, and of course there is the impact such a development would have on an attractive agricultural area. While these issues are initially for consideration by the local planning authority, I intend in due course to call the applications in for public inquiry and my own decision.

'I have been giving serious thought to the most appropriate form of inquiry in this case. It is essential that all the implications of the proposal should be impartially and exhaustively examined. What is the best way of achieving this? One proposal is that we should set up a planning inquiry commission under Section 47 of the Town and Country Planning Act 1971. As you will recall the planning inquiry commission system was introduced into the Planning Acts in the wake of the Roskill Commission on the Third London Airport though it has never been brought into use. It was designed for important proposals which it was felt could not be properly evaluated unless there was a special inquiry, or which involved such unfamiliar technical and scientific aspects that a proper decision could not be arrived at without a special inquiry. However, ten years later we find ourselves in a different situation. At the Windscale Inquiry important changes were made in the scope of the matters open for consideration. That Inquiry demonstrated how the scope of conventional inquiries could be made much broader. Of course I realise that the conclusions reached by the Inspector were not to everyone's satisfaction. But nobody, I believe, is in doubt that the range of the inquiry was exceptionally wide, with the question of need being exhaustively considered and with the Inspector being specifically asked to examine, for example, the national interest, as well as the rightness of the particular site. It is difficult to argue, therefore, that the planning inquiry commission system is uniquely appropriate now for major inquiries.

'There is, however, a further problem with planning inquiry commissions to which I personally do not see a solution. The system envisaged a two-staged procedure, the first being investigatory and the second consisting of one or more public local inquiries. In my view the investigative proceedings are bound to lead the planning inquiry commission to conclusions, by whatever means the proceedings may be conducted. Yet at the second stage, i.e. at the local inquiry, arguments of policy and principle on which they will already have formed a view are bound to be put to them as well as the more local issues, and I do not think that people will feel that they would get a fair hearing. There is no way round this problem. For all these reasons I am not convinced that a planning inquiry commission is the right way to proceed.

'The planning inquiry I envisage on Belvoir would include questions relating to the need for the proposed development and possibilities for alternative locations, as well as important local economic and environmental implications which I understand have been the subject of a joint study by the County Council and the National Coal Board.

'On the organisation of the inquiry, it is already an increasingly common practice in major inquiries for a preliminary meeting to be held to seek agreement between all those concerned on basic facts and to establish areas of disagreement, as well as to draw up an order of business for the inquiry itself. In the present case I propose to ask the inspector to hold such a preliminary meeting, perhaps extended in scope, to identify the main issues on which he considers the inquiry should concentrate, and to indicate the documentation and further work on implications which he would expect to be presented at the inquiry. I hope that this procedure will enable the time at the inquiry to be used in the most profitable manner and will ensure a full and comprehensive examination of all the issues. I shall be making a further announcement in due course about the arrangements.

'But before I leave the NCB's proposal I would like to mention in connection with it another subject in which there has been general interest – the idea of assessing the environmental impact of significant major developments, as part of the planning process. My colleagues and I have considered how best to pursue this. We fully endorse the desirability, as set out in the Thirlwall/Catlow report, which my Department published in 1976, of ensuring careful evaluation of the possible effects of large developments on the environment. All could agree with that, though we must not forget the unacceptable delays and costs of some environmental assessment procedures used in other countries, nor the strong interest we have as a nation in the success of our industrial strategy.

'The Government has already accepted the recommendations of the Leitch Committee, on the assessment of trunk road schemes for future road inquiries. The approach suggested in Thirlwall/Catlow is already being adopted with many other public and private sector projects. We should therefore wish to encourage use of this approach in cases where its use is worthwhile in the circumstances; relevant to the decision; and necessary to the total evaluation of the project along with the industrial, the employment, the social, the health and safety, the land use and the other implications.

'Our feeling therefore is that in selected major cases, involving environmentally sensitive areas or circumstances, a more explicit approach should be pursued. In the selection of such cases, the initiative could come either from the developer or from the planning authority. We should expect that the planning authorities and the public or private developers would agree at as early a stage as possible whether environmental assessment was justified; and if so the form of, and methods of preparing an assessment, including the division of re-sponsibility for carrying out the work. It would be helpful also if detailed consideration could be given to informing all interested parties including the general public of the scope and nature of the analysis to be undertaken. The sensible use of this approach, through the co-operation of all concerned, should I believe improve the practice in handling these relatively few large and significant development proposals.

'This will take time to bring into effect, but at North East Leicestershire I hope that, if necessary, the important environmental considerations that have already been the subject of a joint study by the

County Council and the NCB will be further developed for presentation at the inquiry, perhaps under guidance from the pre-inquiry meeting.

'I now turn to the proposal if and when it comes forward for the first commercial fast breeder nuclear reactor, and here I think the lessons learned from Windscale have much to teach us.

'In handling Windscale I had in mind two major objectives. The first was that I wanted to ensure as thorough an investigation as I could devise. I needed it for my own purposes as Secretary of State, in order to ensure a fully reasoned and informed decision – a decision that was in all the circumstances and with due allowance for human fallibility – right! It was needed also for the reassurance of public opinion, and indeed world opinion, that a thorough investigation had been made. The second objective was to provide for the involvement of Parliament, for it seemed to me wrong to exclude from a decision of such high national importance – one in which the range and depth of the issues was unique and unparalleled, the elected representative of our people. This of course was achieved. There were in fact two debates in the Commons: a full day's general debate on the Inspector's report, and another on the Special Development Order conferring the permission. For the future I am in no sense committed to a Windscale-type procedure, but the same two objectives in my view apply to nuclear issues of the same complexity and importance.

'As I said, the Windscale Inquiry showed that a planning inquiry could range over a very wide field, so that it could take in major national and international issues, as well as questions of need and of environmental concern. And we were all helped by the Royal Commission's 6th Report on nuclear power which provided an informed and detailed background to nuclear development. But that inquiry did not settle whether the particular procedures there adopted were the best in all circumstances. I said at the time that this was new territory, that we were still working out our ideas, and that if we could devise a better procedure we should do so. So we have been asking ourselves whether, should we be faced with another major proposed development in the nuclear energy field we need an arrangement which builds on some of the elements that went into Windscale, but includes also what I hope will be thought other valuable elements suited to the examination of the project concerned.

'My suggestion, which I hope you and others will turn over in your minds, is this. I have already, in the course of the Windscale Inquiry (and in subsequent House of Commons debates), promised a special procedure for public consultation, a wide-ranging investigation going beyond local considerations and – as with Windscale – I am sure that we must involve Parliament in a decision which has the specially wide-ranging and uncertain repercussions attaching to nuclear projects. What I have in mind is a first-stage public examination, by a suitable body such as a Commission or a Committee, outside the inquiry system, to assess the background and the need. The published report of such a body could form a major background document to a subsequent site-specific inquiry. The proposing authority could then be invited to ask the Secretary of State to publish a draft Special Development Order of the kind used for Windscale. This, together with any necessary additional material, would be the subject of a public inquiry with wide terms of reference, held by an inspector and assessors. The report of this inquiry would also be open to public discussion, and the Special Development Order in its final form would be laid before Parliament, becoming subject to debate on a motion to annul.

'To my mind such a procedure would give the most thorough-going investigation possible. In the sophisticated field of nuclear energy it is of the utmost importance to get the answer right. We are a democracy: and we govern by consent. It is our duty to ensure that that consent is justified and to make it possible for the public to feel and to know that the ultimate decisions reached are as wise, fair and acceptable as we are able to make them. This is what we owe to ourselves and that is the responsibility we bear for the future.'

COMMENT

Some brief comments on the ex-Secretary of State's speech are in order in the light of the extensive discussion on planning issues in this report.

First, the fact that change is being contemplated at all is very welcome and indicates an awareness that the planning system is not, at the moment, best suited to dealing efficiently with all the issues placed before it.

Second, the speech is clear in its implications that the Department of the Environment's objections to Planning Inquiry Commissions have been sustained. It will be evident that we disagree entirely with the objections to the introduction of PICs, objections which seem to be clearly based on a fear that the Third London Airport inquiry experience would be repeated. It has already been argued that what caused disquiet at that Inquiry and its subsequent reports, rightly or wrongly, was the embracing by the Commissioners of a specific methodology, cost-benefit analysis. The format of the Inquiry was not at fault and its most valuable attributes – the wider range of expertise available from seven Commissioners and the investigative powers it had – deserve to be emphasised. The reasons given in Mr Shore's speech for objecting to PICs in principle are weak. That it is not *uniquely* suited to major planning inquiries is open to serious question in that it presupposes that a local planning inquiry can efficiently deliberate on issues with national dimensions. Mr Shore is clear that WPI demonstrated that the LPI could be so modified. It has been the message of this book that, if the decision that emerged from the WPI was the right one, it was right *despite* the proceedings and format, not because of it. We repeat one issue alone – the inadequate investigatory powers of an Inspector (or Reporter in Scotland) to support that case. A PIC has such powers. It is not evident that, in the case of Belvoir coalfield, the hoped for NCB/local authority environmental impact statement will be an adequate substitute for a detailed and independent investigation by a PIC.

The second objection is curious and reduces to saying that because a PIC is a two-stage process, with the first dealing with the principle and the second with the local inquiry aspects, the two stages could repeat evidence. This repetition could be costly. Yet, if one consults the Third London Airport inquiry procedure it will be seen that issues of 'need' were distinctly not discussed at the local inquiries and were confined to the major inquiry procedure. Even without this precedent, it is a simple matter to determine terms of reference for local inquiries following a PIC Stage One which exclude the issue of accepting or rejecting the investment *in principle*. It is also clear that Stage 2 of PIC would give rise to fears that Commissioners had already 'formed a view' of site selection at Stage 1. If so, objectors could challenge the legality of Stage 2. Such a problem is easily resolved by making the two stages independent in terms of Commission membership, although this has some undesirable features. This 'legal' fear clearly underlies the ex-SOSE's speech.

The emphasis on the pre-inquiry meeting is however to be thoroughly welcomed and is entirely in accord with the criticisms and suggestions advanced in this report.

Turning to nuclear investments, we have argued that there are reasons why they should be treated as 'sensitive areas'; but it would be unfortunate, in our view, to persist with this view to the detriment of treating other issues, such as LNG plant, as being less sensitive. But the idea of referring the first stage of the CDFR inquiry (it remains a further oddity, and confirmation of our analysis of the semantic confusion within the establishment that Mr Shore continued to speak of CFR 1) to a body outside the planning framework is again totally in keeping with the arguments advanced here. There can be little question that Mr Shore has in mind an existing commission. We have advanced our reasons for saying that no existing commission can meet this function adequately. Further, the idea of involving Parliament directly is again in total keeping with what we have called the 'democratic requirement'.

Under Mr Shore's suggestion, the deliberations of this committee or Commission would constitute Stage One of the proceedings. Then the local authority would seek a Special Development Order for the investment (which tends to assume a single site application, which is not a foregone conclusion for CDFR). The SDO and presumably the report of the Commission/committee would then go to a local inquiry with inspectors and assessors. The report from this stage would then be discussed and a modified SDO would result which would then be presented to Parliament. In this way Mr Shore secures links between Parliament and the planning process just as we have suggested.

As was noted in Chapter 5, however, we would criticise the use of the SDO procedure because it apparently leads to the negation of any rights local inquiry participants may have to raise legal objection to the form or content of the Inspector's Report and the normally subsequent decision by the Minister. It is hoped that the process envisaged by Mr Shore allows for the maintenance of the legal rights currently possessed by inquiry participants.

Some questions remain. It is unclear how the procedure would operate if CDFR is a multi-site application since, technically, it cannot then go to a local public inquiry. It is also not clear what 'status' the Commission/committee report would have at the local inquiry – i.e. whether it is effectively like an environmental impact statement that would form the basis of the inquiry discussion. However, its presence together with a draft SDO could lead to some difficulty in that the draft SDO may itself be seen as some kind of authoritative stamp of recognition on the investment. None the less, the rethinking of procedure is to be welcomed even if its finer detail is not yet established.

Lastly, and of major concern, the speech is clear in its implications that WPI *could* be repeated for other major investments. As argued at considerable length in this book, that would seem to us to be a retrograde and risky step serving only to foster all the consequences that we have described as being the undesirable outcomes of an inefficient procedure.

Appendix 1
Who Was Who at the
Windscale Inquiry

PRO-THORP GROUPS

British Nuclear Fuels Ltd
represented by Lord Silsoe, QC, and Mr G. Bartlett.
BNFL formed 1971. UKAEA own share capital for Secretary of State
for Energy. BNFL's principal activities concerned with provision of
nuclear fuel cycle services.

Witnesses: Mr C. Allday, Managing Director, BNFL, BSc Chemistry. Former employee of UKAEA and ICI.
Discusses: Need for the plant.
Mr A. I. Scott, Company Secretary and Director of Administration. BA Economics. Former UKAEA employee.
Discusses: The works history, planning applications, future developments, employment.
Mr B. F. Warner, BSc, FRIC. Deputy Head of Research and Development, Chartered Chemist. Gives general description and objectives of THORP.
Mr L. P. Shortis, Assistant Director of the Reprocessing Division.
Discusses: Hazards associated with nuclear plant.
Mr J. Doran, BSc, ARIC. Deputy Manager of Windscale and Calder works, Postgraduate diploma, Chemical Engineering.
Discusses: Reprocessing at Windscale.
Mr P. B. Smith, BSc, Mech. Eng. Works Manager responsible for plutonium conversion plants and plutonium fuel plants at Windscale and Calder.
Discusses: Plutonium production and conversion. Dis-

charges to environment, safeguards and security.

Dr. D. W. Clelland, MRICE, FIC. Manager of Research and Development Programme, Reprocessing Division, Risley.

Discusses: Radioactive Waste Disposal.

Mr A. D. W. Corbet, BSc, Mech. Eng., MIME. Chartered Engineer. Senior Technical Officer, Risley.

Discusses: BNFL solidification programme, HARVEST.

Mr P. W. Mummery, MA, FIP. Company Director of Health and Safety and member of the company executive.

Discusses: General health and safety policy at Windscale.

Dr G. B. Schofield, AIMechE. Company Chief Medical Officer. Experience on health aspects of radiation work.

Discusses: Effects of radiation medical surveillance.

Mr S. T. Hermiston, BSc. Senior Health Physicist Windscale and Calder works. Member of Society for Radiological Protection.

Discusses: Radiological protection at the works.

Mr J. K. Donoghue, BSc, FIP. Manager of the Safety Assessment Group in the Reprocessing Division at the Windscale and Calder works. Member of the Society for Radiological Protection.

Discusses: Accidents. How potential for accidents taken into account throughout and after construction of plant.

Mr W. G. Milne, MIME, AMBIM. Chartered Engineer. Engineering Manager (Nuclear Transport), Reprocessing Division, Risley.

Discusses: Transporting of nuclear fuel.

Mr P. W. Wilson, BSc Chemistry, ABIM. Manager of the Plutonium Business Centre, Risley. Co-ordinates all movements of Pu and U for BNFL and UKAEA.

Discusses: Regulations for movement of nuclear fuel.

Mr S. Williamson, MIME. Chartered Engineer. A principal and technical officer with transport section. Ensures BNFL follow regulations for transport of nuclear fuel.

Discusses: Procedures for ensuring safety of transporting radioactive materials.

Dr D. C. Avery, BSc, PhD, FIP. Deputy Managing Director. Responsible for financial, planning activities, involvement with national and international nuclear safeguard arrangements.
Discusses: Non-proliferation Safeguards, Sabotage, Terrorism Security.

Mr M. St John-Hopper, FRICS. Independent witness.
Discusses: Planning aspects of BNFL application.

South of Scotland Electricity Board
represented by Mr D. A. O. Edward, QC, and Mr D. MacFadyen.
SSEB are an autonomous board. Responsible in the South of Scotland District for the complete process of electricity generation, transmission, distribution and sales. Cover an area of 4 million population. SSEB operates Hunterston 'A' Station (Magnox) and Hunterston 'B' Station (AGR). SSEB also intend to build a nuclear power station at Torness. BNFL supply SSEB Stations with new fuel, reprocessing and safe storage of radioactive waste materials. SSEB want BNFL to provide reprocessing services for uranium oxide fuel from Hunterston 'B'. SSEB wish to have 1000 tonnes of oxide fuel reprocessed at Windscale in the first 10 years of operation of the new plant. 30 per cent of SSEB requirements expected to come from nuclear power.
Witness: Mr J. N. Tweedy, SSEB Nuclear Safeguards Engineer. Responsible for advising board on formulation of policy in respect of nuclear safety and nuclear fuel management. Former UKAEA employee. SSEB prefer uranium oxide waste fuel reprocessed rather than stored. Option to build FBR should be kept open.

Central Electricity Generating Board
represented by Mr M. Fitzgerald, Mr Thomas, Deputy Solicitor, CEGB.
Associated with view of BNFL and SSEB. CEGB have an AGR programme. By 1990 a large quantity of irradiated oxide fuel will have arisen from AGR stations. Against long-term storage of irradiated oxide fuel. In favour of reprocessing. Option to build FBR should be kept open.
Witness: Dr J. K. Wright, MA, PhD Natural Sciences, FIP, MIEE. Generation Studies Engineer in Planning Department, Head of Branch, having direct responsibility for advising CEGB on THORP proposals.

Cumbria County Council
represented by MR I. Glidewell, QC, Mr M. Rich.
Cumbria County Council planning committee approved BNFL application. The decision was made with advice from an independent consultant, Professor Fremlin, and after taking into account a considerable body of objections. CCC hopes to secure benefits for the community in employment, training, improvement of infrastructure of area (contributions expected from Central Government and BNFL). Is concerned that monitoring of radioactive discharges into the environment is maintained at a high standard.

Witnesses: Mr T. P. Naylor, MICA. County Councillor CCC and Chairman CCC.

Discusses: Way in which CCC policies developed in dealing with the planning applications submitted by BNFL. Council approves if risks from THORP are acceptable.

Professor J. H. Fremlin, PhD Cambridge (Arts), PhD Birmingham (Science), FIP, FIEE. Member of New York Academy of Sciences. Research in nuclear physics and application of nuclear physics to dentistry and medicine. Chair in Applied Radioactivity, Department of Physics, Birmingham. Latest research in controlled irradiation of human beings for medical diagnostic purposes. Member of CND. Founder member of the BSSRS.

Discusses: Advice to CCC on the effects of the planned expansion of BNFL's operations on the health and safety of the public.

Mr B. Alexander, FICE, FIWES. Assistant Director, Resource Planning, North West Water Authority.

Discusses: There is no ground arising out of provision of water services for refusing BNFL planning permission.

Mr P. Sefton, BA, MEng, MRTPI. Chief Assistant (Transportation) in CCC Department of Highways and Transportation.

Mr A. W. Dalgliesh, MTPI. Assistant County Planning Officer. CCC receives advice on Safety aspects from NII, NRPB, Professor Fremlin and County Planning Officer on safety aspects.

Mr R. D. Donnison, Chartered Surveyor, Chartered Town Planner, Assistant County Planning Officer.

Responsible for development policy including preparation of structure plan.

Discusses: Background to employment and industrial problems and impact BNFL expansion will have locally.

Mr S. H. Murray, farmer, retired barrister, County Councillor, former Chairman of Town and County Planning Committee. At present Chairman of Lake District Special Planning Board.

Discusses: Considerations which the Committee gave to the BNFL applications.

Copeland Borough Council
represented by Mr P. N. Denson.
CBC consists of 48 members. Represents 70,000 people. Area includes large part of Lake District National Park. Associated with the view of CCC on the introduction of new monitoring systems on radioactive releases to the environment, employment, training. Some concern for pressure expansion will put on infrastructure of area. Want improvements in infrastructure partly paid for by Central Government and BNFL.

Witnesses: Mr J. E. Bailey, BA, MRTPI, Planning Officer, CBC.

Discusses: CBC has set out to ensure that improvements to the infrastructure would be made if BNFL application was approved.

Mr J. E. S. Swift, FRICS, MIMunE, Chartered Engineer, Director of Technical and Planning Services, CBC.

Mr R. Battersby, Chief Building Control Officer. Called in to confirm THORP foundations had not been laid.

Electrical, Electronic, Telecommunications and Plumbing Union
represented by MR P. Adams, a National Officer of EETPU Chairman of the Company Joint Industrial Council, a negotiating body on wages and conditions for industrials within BNFL. Appears for workers and Union in the plant. Supports BNFL expansion and Energy Policy laid down by TUC.

Discusses: Need for co-ordinated Energy Policy which would seek to develop all available energy sources. Need to develop methods of energy conservation and improved use of energy. Reprocessing an important part of such a programme. Wants option to build FBR kept open. Supports overseas contracts to offset costs, cheaper services to CEGB, assist balance of payments. Does not want right to strike curbed. Workers will have a responsible attitude towards the plant.

Ridgeway Consultants
represented by Dr K. Little.

Witnesses:　Dr K. Little, MA Oxford, BSc, DPhil, for research on structure of some aluminium and chromium alloys. Has also done medical and pathology research.

Discusses: The biological mechanisms of and possible treatment of some types of radiation injury. Concerned for accurate information on aspects of radiation, especially health.

Dr S. H. H. Bowie, Geologist. Studied uranium geology. Chief Geological Consultant to the UKAEA, 1955–77.

Mr G. H. Greenhalgh, BSc, Chemistry. Former Director of British Nuclear Forum. Former AERE Harwell employee.

Mr S. E. Rippon, BSc Physics. European Editor *Nuclear News*. Freelance technical writer.

Dr J. M. Fletcher, MA Oxford, PhD University of California, FSAL, FIMM. Chartered Engineer. Former UKAEA employee. Working on study and applications of dendrochronology. Chairman of Berkshire Branch of CPRE. Gives evidence on his employment as a Senior Principal Scientific Officer and as Chief Scientific Officer with UKAEA. Supports THORP in principle but with reservations about size of plant.

Mr A. J. MacDonald, student member of Royal Overseas League, Monday Club, United Nations Association.

Discusses: Whether a plutonium economy would mean a 'police state'.

ANTI-THORP GROUPS

Friends of the Earth, London, Ltd
represented by Mr R. Kidwell, QC, Mr O. Thorold.

FOE activities include research, education, publishing, lobbying. Aim to develop and advocate programmes for the conservation and rational use of the earth and its resources. FOE London is part of FOE International. FOE International is accredited to the UN as a Non-governmental Organisation. Granted observer status at UN International Atomic Energy Agency conference on 'Nuclear Power and its Fuel Cycle'. FOE have been involved in British energy and nuclear matters since 1973. Oppose BNFL on grounds that proposed plant is unnecessary, liable to be technical and economic failure, aggravate problems of radioactive waste management, proliferation of nuclear weapons.

Witnesses: Mr G. Leach, BA (also witness for SEI). Senior Fellow of the International Institute for Environment and Development. Directs energy programme for the Institute.

Discusses: The necessity to investigate the most socially, economically and environmentally attractive energy futures before making any irreversible commitment to nuclear power. Questions the assumption that nuclear capacity in the UK must expand substantially and soon in order to help meet a large predicted growth in energy demand by the end of the century.

Mr W. Patterson, MSc, Nuclear Physics. On staff of FOE. Energy specialist. Editor of *Nuclear Times*.

Discusses: The FOE case against THORP generally, i.e. that it is (i) unnecessary (ii) liable to be a technical and economic failure (iii) likely to aggravate waste management problems (iv) liable to undermine international efforts to control proliferation. Alternatives to reprocessing are available.

Dr P. Chapman, Member of OU academic staff. Senior Lecturer Physics, Director of OU Energy Research Group. Former consultant to Energy Technology Support Unit, Harwell, the Programmes Analysis Unit and the Transport and Road Research Laboratory.

Discusses: Energy policy issues; the 'energy gap' and how to fill it. The need for more research and development on non-conventional fuels.

Professor A. Wohlstetter, Professor of Political Science, Chicago. Consultant to the Nuclear Regulatory Commission, USA. Director of Studies for Arms

Control and Disarmament Agency and Defence Nuclear Agency and Defence Advanced Research Projects Agency.

Discusses: Concern with problems of predicting civilian and military technologies and dealing with the uncertainties of such long-range forecasts.

Friends of the Earth, West Cumbria
represented by Mr C. Howarth, BSc, Biochemistry. Formerly employed as Chemical Engineer in oil industry. Involved in production, research and development. At present Head of Biology Department, Whitehaven Grammar School. Member of FOE since 1973. FOE West Cumbria is autonomous, engaged in activities in accordance with general FOE philosophy. Associated with NNC and FOE London. All members of FOE West Cumbria are local people. All witnesses are local people.

Witnesses: Nurse F. Corkhill.

Discusses: Health issues. Lack of overall support of BNFL expansion locally, including Windscale workers.

Mrs J. Norman, local farm worker.

Discusses: Concern with storage of radioactive waste. Worried about apparent lack of concern for safety of residents.

Mrs J. MacLeód, Workington resident.

Discusses: Concern with lack of information. Worried that plant expansion would increase hazards.

Mrs E. M. Jones, writer. Elected FRSL. Member of Religious Society of Friends. Against production of plutonium.

Councillor W. Dixon. Was on the planning committee of CBC when BNFL expansion plans were presented. He suggests the plans were passed without full knowledge of the consequences expansion would bring. Campaigned for the inquiry.

Discusses: Lack of total public support, queries whether expansion is answer to unemployment. Also concerned with management of the works.

Mr D. Bainbridge, Process worker in Health and Safety Department, Decontamination Centre, Windscale. Member of GMWU.

Discusses: His concern with safety aspects of expansion
and safety aspects of the present works.

Society for Environmental Improvement
represented by Mr J. F. Tyme, anti-motorway campaigner.
Inquiry experience. SEI members include scientists and MPs. SEI
concerned in particular with energy matters as they affect the environment in its widest human concept. Also Dr R. Todd and Sir Martin Ryle
on list of witnesses.

The Society for Environmental Improvement represented the
National Centre for Alternative Technology.

National Centre for Alternative Technology
represented by Mr G. Morgan-Grenville and Mr J. F. Tyme.

Witnesses: Professor J. Page, Professor of Building Science. Interest
in energy conservation and solar energy. Member of
International Solar Energy Society.

Discusses: The economic case for significant further
expansion of nuclear power in the UK has not so far
been demonstrated and a more adequate national
energy strategy should be devised that reviews the need
for THORP against progress made in all branches of
energy.

Mr A. Scargill, President, NUM, Yorkshire area.
Yorkshire miners sponsors and co-founders of Energy
2000, an umbrella organisation to co-ordinate the
efforts of various anti-nuclear power groups.

Discusses: Energy demand and the assumed necessity of
extending the nuclear power programme in Britain.
States that there is no need for a nuclear power
programme to meet the UK's energy demand over the
next 50 to 100 years. Alternative fuels can fill the
'energy gap' with the necessary investment; solar
power is the energy of the future. The nuclear power
programme, if allowed to continue, develop and expand, will be a disaster for human life.

Dr P. Musgrove, Chartered Engineer. Research in wind
energy systems.

Discusses: Wind energy systems.

Mr S. H. Salter, Lecturer in Department of Mechanical
Engineering, Edinburgh University. Research in extraction of power from sea waves.

Discusses: Wave power; the UK's ability to afford the deferment of any expansion of the nuclear power programme for two or three years while wave power, as a viable source of energy, is investigated.

Professor E. Wilson, Chartered Civil Engineer, Professor of Hydraulic Engineering, Salford University. Research in tidal energy generation.
Discusses: Tidal energy technology.

Lord Wilson of High Wray, BSc, FICE, FI MechE. Resident of Kendal. Interested in development of hydroelectric power.
Discusses: That hydroelectric power could be better developed in the UK.

Professor D. O. Hall, BSc, PhD, Plant Physiology. Professor of Biology, King's College, London. Chairman of UK section of the International Solar Energy Society. Project Leader, EEC Solar Energy Programme. Interested in bio-fuels.
Discusses: The work going on around the world on bio-fuels and the need to give more serious consideration to the potential contribution which bio-fuels could make to UK energy supplies.

Dr J. E. Randell, Chartered building services technologist, Senior Lecturer, Department of Civil Engineering, Salford University. Involved with energy storage projects. Interested in energy conservation.
Discusses: Energy storage methods for use in houses.

Mr C. M. de Turville, BSc, Physics. Former experimental officer, Harwell, and research officer, Rolls-Royce Nuclear Engineering Lab. At present Research Officer, CEGB, working on environmental problems associated with air and water pollution. Interested in alternative sources of energy, fossil fuel.
Discusses: The extraction of organic fuels from marine sediments.

Mr R. J. Armstrong-Evans. Owns company which constructs small hydroelectric installations. Interested in hydroelectric power and other alternative energy sources.
Discusses: Small hydroelectric systems; alternative power systems.

Sir K. Spencer, Fellow of University College London.

Former FICE, FRAS. Former Chief Scientist, Ministry of Fuel and Power. Member of Council, University of Exeter.
Discusses: Background to nuclear energy. Feels problems associated with nuclear technology should be solved before expansion takes place.
Mr C. H. Armstead, FICE, FI MechE., FIEE, ACGI. Member of 1976-appointed working group of the Watt Committee on Energy to study alternative and renewable energy sources.
Discusses: Possibility of geothermal developments in UK.

Scottish Campaign to Resist the Atomic Menace
represented by Mrs D. M. Paulin, BA (Arts and Commerce), Chairwoman, SW branch.
SCRAM represents people north of the Solway. Support FOE and Isle of Man Energy and National Resources Group. Object to nuclear power on grounds of danger, economics, pollution, ethics. Believe nuclear power to be unnecessary.
Witness: Professor Ivan Tolstoy, scientist and writer (also witness for Windscale Appeal). Favours alternative sources of energy and energy conservation.
Discusses: Disposal of high-level waste; that any further expansion of nuclear facilities must be contingent upon the final solution of the waste disposal problem. Reprocessing makes the management and disposal of nuclear waste more dangerous and reprocessing of foreign fuel would aggravate the British high-level waste problem.

Socialist Environment and Resources Association
represented by Mr M. George, SERA energy group.
SERA concerned with raising broad environmental and resource issues within the various organisations of the labour movement. Membership of 500 trade unionists and Labour Party members, 24 MPs including Secretary of State for the Environment. Explore the various employment issues related to the BNFL expansion. Concerned about the nature of employment. Support FOE on safety/security issues.
Witnesses: Mr S. Sedley, lawyer.
Discusses: Labour Party Conference and TUC against

energy policy SERA advocates. *But* supports SERA's view on employment rights and need for job protection. Concerned with employment issues arising out of expansion plans, i.e. type of employment, alternative employment outside BNFL, employment rights and industrial conflict.

Dr D. A. Elliott, nuclear physicist. Former employee UKAEA and CEGB. At present Lecturer in Technology, OU and Trade Union Liaison Officer for SERA.

Discusses: Nuclear power technology. Windscale expansion. A poor investment in terms of creating jobs.

Mr R. M. Lewis. Degree in Law, MSc Economics and Industrial Relations. Lecturer in Industrial Relations, LSE.

Discusses: Conspiracy and Protection of Property Act. Role of troops. Trends towards authoritarianism Security systems v. employees' rights.

Miss J. Bartlett, BA, Geography. Research Worker, SERA.

Discusses: The economic activity that would be most suitable for the area in terms of employment.

Network for Nuclear Concern
represented by Mr E. Acland, Dr B. Wynne, Mr D. Laxen.
NNC arose specifically out of BNFL's expansion plans. A local group representing the following:

Half Life Lancaster
Friends of the Earth, West Cumbria
Half Life Furness
Half Life West Cumbria
Friends of the Earth South Lakeland
Friends of the Earth Lancaster University
Ambleside Nuclear Concern

Membership also includes individuals and professionally qualified scientists. FOE West Cumbria and NNC organised the petition calling for the public inquiry. Monetary support from 23 UK counties, MPs, organisations, individuals. Associated with TCPA on public health implications. Concerned with public health implications, especially regional environmental aspects.

Witnesses: Mr J. J. Thomson. Born locally. BSc, Biochemistry and

Physiology. Teacher of Biological Science. Experience of fine chemical and antibiotic industries.

Discusses: Concern with control of effluent.

Mr D. P. H. Laxen, MSc, Environmental Science, PhD student. Concerned with atmospheric chemistry. Member of Half Life Lancaster. No formal expertise in nuclear power technology but has looked into energy matters, nuclear power.

Discusses: Concern that too many uncertainties exist about the behaviour of radioactive substances in the environment, pathways to man, ability of BNFL to limit impact of their discharges to the environment.

Dr B. E. Wynne, MA Natural Sciences, PhD Electrochemistry of Intermetallic Alloys, MPhil Edinburgh, Sociology of Science. Lecturer in School of Independent Studies, Lancaster University. Research in fields of science technology policy and their social implications. A local resident.

Discusses: That civil and military uses of nuclear power cannot be divorced, waste disposal, biological hazards. Interest in the way in which decisions are made.

Dr S. Ichikawa, Assistant Professor at the Laboratory of Genetics, Hyoto University. Experience with radiation and genetics.

Discusses: THORP as a part of the general tendency in Japan to export 'pollution'. Concerned with effects of low-level radiation.

Professor E. P. Radford, Professor of Environmental Epidemiology, Pittsburg University. Teacher of radiation biology. Has done research in various aspects of radiation exposure to lung cancer in man. Appears as a private citizen.

Discusses: Medical public health implications of the operation of the Windscale plant on Windscale workers, local inhabitants, those exposed to effluent from the plant.

Town and Country Planning Association
represented by Sir F. Layfield, QC, Mr R. Barratt.
TCPA is a limited company, registered as a charity, founded in 1899.
Concerned with land planning and the environment. Concern for needs

of individual households, family, quality of physical environment. TCPA prefer open decision-making at national level on such an important issue as the BNFL expansion. Concern for effective planning. Would have preferred PIC. Associated with Half Life Cumbria and NNC. Joined local groups in pressing for public inquiry.

Witnesses: Dr D. G. Hall, MRJPI, Director of TCPA. Introduces TCPA.

           Dr A. M. Stewart, and colleague Mr G. Kneale. Dr Stewart was Director of Oxford Survey of Childhood Cancers. Concerned with low-level radiation and cancer.

           Professor J. Rotblat, Emeritus Professor of Physics, University of London. Worked on development and construction of first atom bomb. Former consultant to AERE Harwell and Research Laboratory of Associated Electrical Industries, Aldermaston. Founder of Pugwash conferences on Science and World Affairs. Former member of governing body of the Stockholm International Peace Research Institute.

Discusses: Concern with nuclear proliferation.

Professor P. R. Odell, Professor of Economic Geography of the Netherlands School of Economics, Erasmus University, Rotterdam. Consultant to the UK Department of Energy in field of the economics of oil and gas development and production (1978).

Discusses: Energy demand and ways of meeting that demand. Expansion of nuclear reprocessing facilities in UK for use by other countries may not be advantageous in economic terms.

Mr T. B. Stoel, Jr. Attorney employed by NRDC. Experience and involvement with process of environmental impact assessment in USA.

Discusses: Proposed expansion of BNFL plant should be subject to EIA procedures.

Professor R. Ellis, Professor of Medical Physics, Leeds University. Interest in radiation protection and health physics.

Discusses: Need to reduce the occupational exposure of the workers in the present reprocessing facility. Also need to limit discharges into Irish Sea.

Mr C. G. Thirwall, planning expert. Has been engaged in

a study of Environmental Impact Analysis which was commissioned jointly by the Secretaries of State for the Environment, and for Scotland and Wales.

Discusses: Potential environmental impacts other than the impacts on health and safety.

Political Ecology Research Group
represented by Mr P. Taylor.
PERG is an association of research workers from various scientific disciplines. Based in Oxford, includes members of Oxford University and Oxford Polytechnic. Founded autumn 1976. Emphasis on research, drawing together of disparate fields of study. Not a campaign group. Associated with NNC on low-level liquid waste discharges. Aids other groups of objectors to process information. PERG state there is insufficient information made available to the public. Want a safety study made and produced for open public debate. Suggest a survey on the behaviour of plutonium in the environment. Concerned about the possibility of risks to the public from Windscale plant.

Witnesses: Mr P. Taylor, Degree in Natural Sciences, Oxford. Co-ordinator of PERG. Lecturer in Biology and Geological Sciences, and Environmental Sciences. Specialist in ecology, genetics, animal behaviour. Introduces PERG's arguments and views on nuclear technology and how it is viewed by supporters and objectors.

Dr G. R. Thompson, BSc, BMech Eng. A freelance engineer with an interest in natural energy systems and resource conservation. Has done research on nuclear fusion power with support from UKAEA.

Discusses: Concern with safety and accident analysis for THORP.

The Windscale Appeal
represented by Mr D. Widdicombe, QC, Mr A. Alesbury.
Windscale Appeal made up of the following groups:
Concern Against Nuclear Technology Organisation
The Conservation Society
Cornwall Nuclear Alarm
The Ecologist Magazine
The Ecology Party
Greenpeace (London)

Irish Conservation Society
Society for Environmental Improvement
Wexford Nuclear Safety Association

Witnesses: Mrs I. Coates represents Windscale Appeal. Former national chairwoman of Conservation Society. At present the convenor of the Land Use and Planning working party of the Conservation Society and their statutory objector at this inquiry.

Dr C. B. Sweet, Senior Lecturer in Economics, Polytechnic of South Bank.

Discusses: The economics of nuclear power and electricity generation. States that BNFL failed to make a proper economic assessment of THORP.

Dr B. Shorthouse, PhD Cambridge, Chemical Engineering. Former employee EERPG and CEGB. At present with OU also 'on loan' to HEMI as Research Director. Director of two companies in fields of innovative technology and product development.

Discusses: Possibly wiser and safer to have several intermediate steps in the development of oxide fuel reprocessing.

Professor G. Atherly, Professor and Head of Department of Safety and Hygiene, University of Aston. Had various posts in medicine and commerce, occupational health, pure and applied physics.

Discusses: BNFL expansion should not be allowed until special provisions for health and safety are made.

Dr C. M. H. Pedler, author.

Discusses: Existing security arrangements not sufficient to withstand terrorist attack (not a security expert).

Dr C. Wakstein, PhD Nuclear Engineering. Interested in engineering and design safety.

Discusses: Attitude to safety in the nuclear industry and BNFL inadequate for the highly advanced and hazardous technology involved.

Mr E. D. Goldsmith, MA Oxford, PPE, Editor of *The New Ecologist*.

Discusses: Need for a system of production which is not wasteful of energy resources and has a minimum

impact on environment. In favour of low technology
and a highly decentralised society.

Dr J. Davoll, PhD Organic Chemistry, Cambridge.
Former Assistant Director of Chemical Research for
pharmaceutical manufacturer. At present Director of
the Conservation Society Ltd.

Discusses: Concern for balance of nature. Has studied
social, political, economic policies in relation to the
environment.

Dr R. E. Blackith, BSc, Chemistry, PhD for application
of mathematical methods to biological problems. FIS.
Fellow, Trinity College Dublin. Lecturer in Zoology.
Represents Irish Conservation Society and the
Wexford Nuclear Safety Association.

Discusses: That there are inadequately examined assumptions which seem to underlie the proposals for BNFL
expansion.

Mr M. Jenkins, energy expert.

Discusses: Making use of both heat and electricity
produced in power stations in order to reduce amount
of electricity needed, thereby reducing need to build
new power stations.

Isle of Man Government Board
represented by Mr G. L. S. Dobry, CBE, QC, Mr J. C. Harper.
Witnesses: Mr P. H. Newbold, Spokesman for Isle of Man.

Discusses: IOM concern with effect of discharges of
radioactive waste into Irish Sea. Concerned with effect
of discharges on the environment generally. Need for
IOM to be kept informed of technological controls and
safety measures. IOM want to be assured Windscale is
best site for THORP and that economic advantages to
UK outweigh drawbacks.

Dr V. T. Bowen, Geochemist. Senior Scientist, Wood
Hole Oceanographic Institution, Mass., USA. Interest
in biogeochemistry of artificial radio-nuclides.

Discusses: The discharge of transuranic nuclides to the
Irish Sea. Concerned about acquisition of the maximum amount of information available with relation
to the behaviour and effects of the radionuclides
released from Windscale. Presents arguments for a

different philosophy and practice of monitoring and control.

Mr R. B. M. Quale, Clerk of Tynwald, Secretary of House of Keys, Solicitor of Supreme Court of Judicature.
Discusses: Press, local environmental groups are concerned about pollution of Irish Sea.

Mr A. B. Bowers, Senior Lecturer in Marine Biology, Liverpool University. Works at University Marine Biology Station, Port Erin, IOM. Station has a statutory obligation to give advice on fishery and marine environmental matters to IOM government. Gives factual evidence on marine fishery matters. No special knowledge of effect of Windscale radioactive discharges on marine organisms or on the people who eat fish caught in Irish Sea.

Professor G. W. Ashworth, FRIBA, FRTPI Professor of Urban Environmental Studies, Salford University. Member of NW Economic Council and Countryside Commissioner.
Discusses: The strategy of site selection.

National Peace Council
represented by Mrs S. M. Oakes, General Secretary of the NPC.
Member of the NGOSCD in Geneva, which has an official relationship with the UN.
NPC has 74 affiliated organisations. Founded in 1908. Affiliated to the IPB in Geneva. NPC objections to BNFL expansion are in connection with the production of plutonium and its use in nuclear weapons. Cannot separate nuclear power production from nuclear weapon production.

Windscale Inquiry Equal Rights Committee
represented by Mr J. Urquhart.
WIERC was formed after the announcement of the public inquiry. Its aim is to ensure the widest possible discussion of the issues surrounding the Inquiry and the decision that is necessary on THORP.
Witnesses: Mr J. Urquhart, BSc, Secretary of WIERC. Former statistical adviser to the LMRU, Cambridge. Former Scientific Officer, Marine Laboratory, Aberdeen. At present Head of Aquisitions, Newcastle University

Library. Interested in information aspects in technological decision-making and the eventual impact on society.

Discusses: Concern with risks and uncertainties surrounding THORP.

Professor W. L. Boeck, Professor of Physics, Niagara University, New York. Presents his own views. Chairman of the Krypton-85 Working Group of the International Commission on Atmospheric Electricity.

Discusses: The plan to dispose of large quantities of radioactive gas by releasing it into the atmosphere. Considers the implications of massive releases of Kr-85 and alternatives to a total relase of Kr-85.

Friends of the Lake District
Cumbrian Naturalist Trust
Mr G. Berry, Consultant Secretary to FLD, represents FLD and CNT. Concerned with visual impact of development. CNT established 1962 as 'Lake District Naturalists Trust'. A registered charity. Affiliated to the Society for the Promotion of Nature Conservation. CNT has 2000 members. Owns, manages or has access agreements over 24 nature reserves in the country. Concerned with hazards to wildlife and the natural environment. Concerned with direct and indirect effects of industrial development and any special effects in proposed BNFL development arising from radioactive releases.

Witness: Dr G. Halliday, MA, PhD. Council Member of Lake District National Trust. Reads proof of Miss Ketchen, the Conservation Officer of CNT.

Council for the Protection of Rural England
represented by Mr M. Kimber. Employed by Lancashire Branch of CPRE but also represents the national organisation.

Society of Friends Keswick
represented by Mrs J. Spearing, housewife. Lives in Keswick area. SoFK has a large body of Quaker support.

Discusses: Responsibility to future generations. Windscale expansion can have harmful effect on society at large and on health of present and future generations.

Council for Science and Society
represented by Mr D. Widdicombe, QC. A registered charity. Engaged

in study of the social consequences of science and technology with communication of the results to the public.

Justice
represented by Mr D. Widdicombe, QC. Member of the Council and Executive Committee of Justice.
Organisation founded 1957. 1500 Members. Justice is an all-party association of lawyers. Concerned to uphold the principles of the rule of law, assist the administration of justice, preserve fundamental liberties of the individual. Justice is the British section of ICJ. On BNFL expansion, Justice is concerned that UK may take a leading step to the modification or loss of its free institutions and fundamental liberties.

Witness: Mr P. Sieghart, Joint Chairman of Executive Committee of Justice. Governor of BIHR on Executive Committee of International Commission of Jurists. Former statistician and electronics engineer.

Discusses: Concern with possible consequences of an expanded nuclear programme on rights of UK citizens and effects on legal system.

National Council for Civil Liberties
represented by Mr G. Robertson, Mr L. Blom-Cooper, Mr S. J. Irwin. NCCL experienced in developing the defence of individual freedom. Believe progress towards more civilised society will be arrested by security measures which will flow from increased storage and transport of plutonium if BNFL expansion allowed. NCCL concerned over lack of information on nuclear security involved in vetting and surveillance of Windscale personnel and movement of dangerous substances.

Witness: Mr R. Grove-White, Oxford graduate, Assistant Secretary NCCL.

Discusses: Conflict between imperatives of security and those of openness. Concerned about secrecy of security measures surrounding nuclear power.

British Council of Churches
represented by Dr D. Gosling, degrees in Theology, Physics, Nuclear Physics and Religious Studies.
Objections to THORP because (*a*) experts disagree on technical issues; (*b*) problems of long-term method of waste disposal; (*c*) responsibility and risk passed on to descendants; (*d*) plutonium production incurs unacceptable limitations on civil liberty and risk of proliferation; (*e*)

foreclosing of alternative and less dangerous options by decision to expand a part of nuclear energy process which gives prominence to use of plutonium; (*f*) relative insignificance of argument based on foreign earnings and employment.

Witnesses: Mr G. S. Ecclestone, Secretary of Board for Social Responsibility of Church of England. BCC wants to see responsible stewardship of resources and natural environment, calls for a comprehensive long-term British energy policy. BNFL application raises moral issues on nature of society, responsibility for future, risks we are entitled to take. Importance of making explicit the values which underlie judgements.

Rev. A. J. Postlethwaite, Vicar of Whitehaven.

Discusses: Local unease about Windscale plant. Need for better PR between BNFL and local people. Need for independent advice on radiological matters for local authorities, groups and individuals in the community.

Lancashire and Western Fisheries Joint Committee
represented by Professor W. T. W. Potts.
LWFJC is a public body set up to protect and develop inshore sea fisheries from Haverigg Point, Cumbria, to Cemnaes Head, Pembrokeshire. Committee consists of councillors from the local authorities concerned, water authorities, inshore fishermen, representatives of the wholesale and retail fish trade, scientists and naturalists appointed by MAFF. LWFJC concerned with the unsatisfactory state of the effluent from the existing plant. Has recommendations to make as to the conditions which should be imposed before any further development of the Windscale plant takes place.

Individual Objectors

Mrs U. F. Wadsworth, lives in Bradford, housewife. Anti-nuclear energy. Wants adequate information given to public, democratic decision-making; health, environment.

Mr G. Hatton, lives in Rosendale, Lancashire. Housing Officer. Concerned about safety aspects of nuclear technology.

Mr Stredder, one-man street theatre. Concerned with pollution, safety aspects. Wants a full public debate.

Dr J. K. Spearing, FLSL, FIB, husband of Mrs J. Spearing of SoFK. Concerned with long-term effects of routine radioactive emission.

Mrs Dalton, local resident. States BNFL have sent unreprocessed fuel to Australia where it has been buried.

Mr Miller, ex-employee Windscale Works. Concerned about plutonium contamination.

Mr A. Hillier-Fry, local resident. Concerned for efficient use of energy, energy conservation. In favour of 'soft energy technologies'.

Mr R. Tosswill, local resident. Motor mechanic. Fears catastrophic release of radioactive material to the atmosphere.

Mrs J. M. Henderson, resident of Cockermouth. Concerned with 'quality of life'. Concerned about pollution. Nuclear energy not an economic necessity. In favour of alternative sources of energy.

Mrs A. Dudman, MA Oxford. Teacher in Whitehaven. Member of FOE West Cumbria. Concerned about FBR development if THORP plans approved.

Mrs A. Tremlett, teacher, commercial subjects. Concerned with geographical location of THORP. Proposed site of plant has potential for a natural disaster.

Mrs D. Richardson, local resident. Housewife. Against storing atomic waste because of problems left to future generations. States local people are against expansion. More information needed. Local people ignorant of dangers.

Mr P. B. Chivall. Believes THORP to be unnecessary, uneconomic. Concerned for danger to future generations. Pollution. Threat of plutonium production leading to breakdown of national and international peace.

Mrs M. S. K. Higham, member of Driggecarelton Parish Council. Former UKAEA employee. States local people and BNFL employees at Windscale concerned about health hazards. Research needed on waste disposal. Amount of radioactive substances handled at Windscale works should not be increased.

Miss B. Fish, an impromptu objector. Does not see need for nuclear fuel at all. Mr N. Beckett and Mr P. Scott called to give their views.

Mr W. C. Robertson, Chartered Mechanical Electrical Engineer. Former UKAEA employee at Windscale and Springfield works. Wants nuclear development kept at a minimum, to meet essential needs of UK. Concerned with safety aspects of plant.

Mr I. C. Holden states that bureaucracy is in way of individual objectors and that specialists dominate the inquiry while lay people are ignored. Member of the Northern Friends Peace Board of the Society of Friends. Concerned for life and liberty of individuals.

Mrs S. H. Sly, local resident. Teacher. States local people are against THORP. Concerned with safety aspects, and with fact that experts cannot agree.

Mrs R. M. Jones, BA. Resident of Ilkley, West Yorkshire. Housewife. Member of Half Life. Concerned about waste disposal seepages, labour relations, management at plant. Prefers alternative technology. Need for more public discussion information, research into attitudes of public.

Dr T. B. Cochran, Staff Scientist at the Natural Resources Defense Council, Washington, an environmental group predominantly consisting of lawyers, some scientists. Speaks for NRDC and UCS, both USA groups.

Discusses: Concern with escalation of nuclear reprocessing units. Production of plutonium could lead to nuclear proliferation.

Lake District Special Planning Board
represented by Mr J. R. Robinson.
LDSPB responsible for administration of Planning Acts and associated legislation within the Lake District National Park. 27 members, 18 appointed by CCC, 9 by Secretary of State. LDSPB consulted by CCC about BNFL proposals. Board not pro or anti BNFL proposals but want certain safeguards for their area if BNFL proposals are approved, i.e. that transport would not affect the National Park, in the building of new houses care would be taken to preserve the character of the villages, that water supplies would be adequately dealt with.

Witness: Mr R. B. Baynes, MA. Diploma in Town Planning. RTPI. Chief Planning Officer to the Lake District Special Planning Board.

GOVERNMENT DEPARTMENTS AND ORGANISATIONS

Department of Energy
represented by Mr T. P. Jones, Mr C. Herzig.

Mr T. P. Jones, Deputy Secretary with responsibility for co-ordination of energy policy. Describes Department's task in formulating energy policy and the need for nuclear power generation in the UK. Need to retain FBR option.

Mr C. Herzig, Head of Atomic Energy Division.

Discusses: department's specific responsibility for nuclear power and deals with related matters such as reprocessing and non-proliferation, but approves THORP (subject to outcome of Inquiry) and reprocessing for overseas customers under certain conditions. Government consults NII on safety and protection of environment, also Department of Environment, MAFF.

Department of the Environment
represented by Mr J. R. Niven, Mr B. Hookway.

Mr J. R. Niven, an Under-Secretary. Head of the Administrative Directorate of Noise, Clean Air and Wastes. Directorate concerned with policy and administering of the Radioactive Substances Act, 1960. Secretary of State Environment and MAFF jointly responsible for administering the Act. Also concerned with responsibilities connected with waste management policy.

Mr B. Hookway, Senior Radiochemical Inspector in the Department of the Environment. Radiochemical Inspectorate advises on technical aspects of disposal of radioactive wastes.

Ministry of Agriculture, Fisheries and Food
represented by Mr W. R. Small, Dr N. T. Mitchell, Mr J. A. Carr.

Mr W. R. Small, responsible for one of MAFF's fishery

divisions. Deals with waste disposal. Joint responsibility with Food Science Division for authorisations of radioactive discharges into sea. Responsible for authorising monitoring of effects of discharges into the sea.

Dr N. T. Mitchell, BSc. Special honours Chemistry, PhD. Physical and radiochemical research. Chartered Chemist, FRIC. Head of FRL. Experience in radiological control, water pollution research. Engaged on research into environmental behaviour of radioactivity and application of radioactive traces.

Mr J. A. Carr. Involved with technical aspects of the effects of radioactive discharges on agriculture and the food chain.

Department of Transport
represented by Dr E. J. Wilson, Mr J. B. Liddle.

Dr E. J. Wilson, PhD Radiochemistry. Head of Dangerous Goods Branch. Deals with applications from people who want to transport radioactive materials. Concerned with package design and regulations.

Mr J. B. Liddle, BSc, Civil Engineering, Chartered Engineer, MICE. Senior Engineer. Serves under Regional Controller Roads and Transportation for Northern Region of England, to whom he is responsible for transportation matters generally within Cumbria. Particularly concerned with trunk roads. No objections to expansion of Windscale facility if BNFL improve infrastructure as they state.

The Uranium Institute
represented by Mr T. Price.
The Uranium Institute is an international association of producers and consumers of uranium and other organisations concerned with trade in uranium and other nuclear fuels. It is a non-profit-making company incorporated under British law. The UI is put under government bodies because their membership comprises AEA. International commercial companies are also represented in the Institute.

Mr T. Price, MA, Secretary General.
Discusses: Future uranium supplies.

Atomic Energy Authority
represented by Professor F. Farmer, Dr R. H. Flowers.
    Professor F. Farmer, OBE, FIP. Visiting Professor,
    Imperial College; Safety Adviser, UKAEA. Member
    of Major Hazards Committee and Advisory
    Committee on Nuclear Safety. Offers advice to all
    industry on matters concerning safety reliability
    assessments.
    Dr R. H. Flowers, BSc, PhD Chemistry. Head of
    Chemical Technology Division, UKAEA. Requested
    by Secretary of the Inquiry to make a study of zircaloy
    and stainless steel cladding during storage. Gives
    results of study to the Inquiry.

National Radiological Protection Board
A statutory corporation created by Radiological Protection Act, 1970.
Board consists of a Chairman and 7–9 members appointed by the four
Health Ministers of the UK. Made up mainly of scientific and technical
members. Dr A. S. McLean, Director of NRPB. Members include Sir
Edward Pochin. Functions of NRPB: (1) to advance the acquisition of
knowledge about the protection of mankind from radiation hazards; (2)
provision of information and advice to persons with responsibilities in
UK in relation to the protection from radiation hazards of either the
whole of the community or a certain section of it. NRPB has a formal
obligation to advise the Government on safety standards. 1971 assumed
responsibility for Radiation Protection Service and activities of
Radiological Protection Division of the Central Health and Safety
Branch of UKAEA.
Witnesses: Dr A. S. McLean, Director of NRPB.
        Dr D. W. Dolphin, Assistant Director with special
        responsibility for Research and Development.
        Mr F. Morley, Assistant Director with special responsi-
        bility for Nuclear Assessment.
        Mr G. A. Webb gives evidence on dose levels.
        Miss P. M. Bryant gives evidence on environmental
        monitoring.
        Mr M. O'Riordan, PSO. Reports on survey carried out
        on airborne concentrations of Pu 239 and Pu 240 and
        of Am 241 out-of-doors in Ravenglass.
        Mr K. B. Shaw, PSA, NRPB Harwell. Group Leader in
        Nuclear Assessments Department. Confirms interim

report given by M. O'Riordan, i.e. no public health danger in Ravenglass from airborne Pu and Am during time of study.

Nuclear Installations Inspectorate

NII is part of the Health and Safety Executive, which is a Government Agency established in 1974 by the Health and Safety at Work Act. HSE takes policy instructions from the Health and Safety commission. Commission reports to the relevant Secretary of State (depending on the problem). Represented by Mr J. Dunster and Mr F. R. Charlesworth.

Mr J. Dunster, Deputy Director General Health and Safety Executive.

Mr F. R. Charlesworth, Senior Assistant Chief Inspector responsible for Branch 3 NII, which deals with BNFL's chemical plant at Windscale. Branch 3 responsible for implementation of licensing at Windscale. NII check design and safety of plant from conception, through various stages of development. Use specialist consultants. Discussions with BNFL on design principles of THORP already under way. Authorisation to construct must be given by NII. So far NII see no reason to oppose THORP on safety grounds.

Durham County Council

was represented at the Windscale Inquiry but did not give evidence.

Appendix 2
Summary of Written and
Parliamentary Protests
about the WPI Report

SOURCE	NATURE OF MAIN PROTEST
Professor E. Radford, University of Pittsburgh	1. Mr Justice Parker 'downgraded' testimony on low-level radiation due to inability of witness physically to produce four studies partly relied upon to substantiate case for arguing that ICRP radiation limits are at least 20 times too high.
	2. One of the assessors, Sir E. Pochin, possessed all the 'missing' documents.
	3. Sir E. Pochin was himself a member of ICRP, whose standards and 'up-to-dateness' was being criticised.
	(See *Times*, 7 April 1978; and personal correspondence.)
Network for Nuclear Concern	1. Support for Professor E. Radford's complaint (Radford was NNC witness).
	2. WI Report failed to mention disparity between US radiation standards and those in UK, the former being '20 to 60 times stricter'.
	3. WI Report does not mention US system, advocated by NNC, of setting radiation control costs against implied increase in electricity prices.
	4. WI Report alleges NNC said there was 'cause for alarm' about exposure of Ravenglass residents to ambient Pu and Am. (See WI Report, para. 10.83.) NNC stated there were grounds for 'uncertainty'. The

phrase 'cause for alarm' was never used or implied.

5. WI Report (para. 10.83) refers to NNC's advocacy of a five-year monitoring programme at Ravenglass *before* any assurances are given to its inhabitants. No such advocacy was suggested. The WI Report also implicates a second NNC witness in this view, but this witness did not make any such statement.

6. WI Report (para. 10.84) implies NNC witness advocated a monitoring programme which 'would require samplers to be placed almost everywhere'. No such statement was made.

(See letter by NNC to Secretary of State for the Environment, 22 March 1978.)

Town and Country Planning Association

1. Existing oxide reprocessing plant B204/B205 will be available 1979–81. B205 can handle existing and planned AGR spent fuels 1984 onwards, assuming no foreign contracts (Report, para. 2.33). THORP is not needed, given the reduced ordering programme for AGRs compared to that envisaged in WI Report. The WI Report therefore exaggerates 'need' for THORP.

2. Comparison of ICRP and USA radiation limits ignored despite debate at WPI. (See also NNC, 2. above.)

3. THORP is unlikely to be commercially viable, and there was no cross-examination on BNFL's cost data.

4. Unless international safeguards on proliferation are improved (Mr Justice Parker recommended they *should* be – WI Report, para. 6.6), there is a risk of proliferation. But Parker denies a capacity to assess risk (para. 6.33).

5. National energy policy largely ignored in WI Report despite 10 days of submissions on the subject.

6. Mr Justice Parker bases his argument for seeing no problem in scaling up a 1/5000 plant on 'faith' in the experience of BNFL engineers.

7. Mr Justice Parker based much judgement on the 'impressiveness' of witnesses (to him) and not on what they said.

8. Mr Justice Parker's view that by *refusing* to reprocess the UK would be in breach of the non-proliferation treaty is a misinterpretation of that treaty. His view is that the NPT is a 'bargain' between nuclear states and non-nuclear-weapons states such that the former aid the latter to secure civilian nuclear power provided the latter do not develop military capability. This 'must surely have included the development of reprocessing. . . .' (WI Report, para. 6.16). Japan is given as an example – the UK has no obligation to reprocess Japanese waste under the NPT. Even if reprocessed it would require return to the USA for enrichment and Pu fabrication. The USA could well refuse to do this in which case TCPA fears the direct supply of Pu to Japan, in breach of the NPT. This issue not openly dealt with in WPI Report.

9. Mr Justice Parker's view that THORP would *deter* other countries from establishing reprocessing plant rests on the idea that those currently *with* nuclear capacity can influence those *without* it to develop civilian nuclear power in such a way that they will not construct reprocessing plant. TCPA argue this is a dangerous argument and that the UK should follow the US lead in *abandoning* reprocessing and persuading all others to do so as well.

10. Mr Justice Parker should at least have recommended delay until INFCEP is completed. His view is that if INFCEP comes out

against reprocessing it will merely have lost some expenditure. TCPA argue that, on the contrary, approval of THORP will encourage the French and Germans to reprocess.

11. Mr Justice Parker's comparison of the environmental aspects of coal and nuclear power is misleading in that no objector advocated expansion of coal-based power without technological change to improve environmental impacts.

12. Mr Justice Parker believes BNFL will achieve routine discharges from THORP below basic limits despite fact that they have failed to do so in the past.

13. Mr Justice Parker's dismissal of the Radford and Stewart evidence is out of keeping with his acceptance of the evidence of others on the basis of their experience.

14. Mr Justice Parker's *certainty* about the absence of risk from existing discharges in the Ravenglass area is out of keeping with the limited monitoring there and the expressed doubts of Professor Radford about Pu concentrations in silt.

15. Mr Justice Parker failed to acknowledge the argument at WPI that imposed risk and voluntarily accepted risk are not perceived by the public as of equal status.

16. The storage option was dismissed on grounds of cost and uncertainty, even though the tests on AGR fuel pins were on fuel that had been held in untreated water.

17. Mr Justice Parker bases his belief in the success of vitrification on statements that it will succeed, whereas no full-scale plant yet exists.

18. Mr Justice Parker was wrong to say (WI Report, para. 14.9) that an environmental impact statement would have added nothing to existing documentation. In TCPA's view

the documentation presented was inadequate.

(See TCPA, *The Windscale Inquiry – A Critical Assessment*, March 1978.)

Friends of the Earth

1. Long-term storage of AGR stainless steel fuel pins was inadequately tested since the Report failed to mention that the water storage facilities had not been chemically controlled, as FOE had advocated. (See also TCPA, 16. above.)

2. The Report fails to evaluate, *technically*, the option of gas-cooled storage facilities. FOE had argued that dry storage facilities could be developed in 7–9 years, but the WI Report (para. 17.2) states that such a route would be 'a costly and lengthy process'.

3. Investing in storage techniques is no more uncertain or costly than investing in a reprocessing plant on a scale for which no successful precedent exists.

4. The WI Report is 'asymmetric' – it accepts BNFL's assurances on what is technically possible and rejects other evidence as resting on unproven technology.

5. If THORP fails it will have precluded storage as an option since fuel elements will no longer be in an intact, storable state.

6. Reprocessing does not reduce the amount of waste to be disposed of significantly and increases some of the problems of waste management. Stored fuels are *less* radioactive and hence, even if reprocessing is to occur, delay can reduce the disposal problem. This point is inadequately dealt with in the WI Report.

7. Pu separation is not sensible in the absence of fast reactors and the Report's reference to its use in thermal reactors is in contrast to what 'all witnesses, BNFL included' proposed or expect. The WI Report therefore presupposes a programme of fast reactors

which the Inquiry was meant not to comment on.

8. If, on the other hand, a reactor programme does not exist on a scale sufficient to use Pu, then the Pu inventory is *not* reduced as the WI Report, repeatedly, argues it is.

9. However 'burned', the Pu finally to be disposed of under the reprocessing route is, at best, 1/10th of that under a storage route, not 1/1000th as the WI Report claims. Mr Justice Parker ignores what happens to the recycled Pu.

10. The Report fails to note that leaching from glassified waste is no different from that from solid spent fuel, although it does note (WI Report: para. 8.27) that glassification reduces the surface area for waste disposal. While the 'possibility' of oxidation of spent fuel is referred to, the 'possibility' of vitrified waste becoming crystalline is ignored.

11. Having denied that he can judge between energy forecasts or make them himself, Mr Justice Parker then proceeds on the basis of accepted official forecasts.

12. Mr Justice Parker reports Dr Chapman, a FOE witness, as supporting 'nuclear power and even reprocessing' (WI Report, para. 13.7) when the purpose of Chapman's evidence was to show that even in a nuclear future reprocessing is not necessary.

13. Not reprocessing is *not* equivalent to 'throwing away' resources (WI Report, para. 17.2) since storage permits retrieval.

14. Mr Justice Parker nowhere queries BNFL's estimates of storage costs extrapolated to 2036 and on which technology BNFL has no long-term capability. Having initially doubted the economics of THORP, he then accepts it will be economic.

15. The fact that 'civil' Pu has been exported from the USA in the past is not an instance

of the trust that can be placed in such international transfers but (*a*) an instance of the dangers of such tranfers, and (*b*) a reflection of the fact that they took place at a time when 'everyone' thought civilian Pu could not be used to manufacture a weapon. Issue (*b*) was shown to be wrong only in 1977.

16. Mr Justice Parker interpreted the NPT as 'obliging' the UK to reprocess foreign spent fuel. This interpretation is based on a false paraphrase of the relevant NPT Article. In any event, no such obligation is accepted by the UK Government. (See also TCPA, 8. above.)

17. In any event, Article IV of the NPT is subservient to Articles I and II which require nuclear states *not* to assist, in any way, the construction or acquisition of nuclear weapons by any non-weapons state.

18. Mr Justice Parker misinterprets US policy, which is clearly *against* the building of 'more solvent reprocessing plants at this time'.

19. Foreign countries are unlikely to assent to dependence on the UK as a source of plutonium. Moreover, approval of THORP merely adds a 'label of legitimacy' to foreign reprocessing plant.

20. Mr Justice Parker's view that, even if there is a proliferation risk, it will not occur for 10 years (WI Report, paras. 6.23; 17.6) overlooks disguised reprocessing plant not designed for civil purposes and which can be built much more quickly.

21. Mr Justice Parker's reference to 'technical fixes' such as the irradiation of fuel to prevent proliferation is slight, given that such fixes are at least complex and expensive and the problems are probably 'virtually insoluble'.

22. Despite reference to 'technical fixes' and despite dismissing proliferation objections

to THORP (WI Report, para. 17.6) only to have previously declared *he* would not assess them (para. 6.33), not *one* of the Report's recommendations relates to international aspects of WPI or to the said technical fixes.

(See Friends of the Earth, 'The Parker Inquiry', document presented to the Secretary of State for the Environment, 28 April 1978; also C. Conroy, *What Choice Windscale?* (FOE, Conservation Society, 1978).

Political Ecology Research Group

1. The Report fails to consider adequately the hazards of loss of cooling in a tank containing highly active waste (HAW). Such an accident can be simulated using a computer programme TIRION, possessed by BNFL. Only limited runs of the programme were carried out and Mr Justice Parker refused to instruct BNFL to carry them further to consider prolonged releases (beyond 24 hours). PERG claim such prolonged releases could (*a*) occur due to strike or contamination of the area where repairs must occur, (*b*) result in early deaths in the affected sector of 95–4450 persons (depending on weather, 1850–11,300 fatal cancers and lung morbidity of 400–24,100 persons). The WI Report fails to convey this hazard possibility which is alarmingly greater than anything envisaged in BNFL's evidence.

(See PERG, Oxford Report No. 4, *The Windscale Inquiry and Safety Assessment*, August 1978. PERG is documenting a full critique of the WI Report and WPI – these were not available at the time of writing.)

Windscale Appeal

1. Witness Tolstoy misconstrued in WI Report. Mr Justice Parker says 'final effect' of Tolstoy's evidence was to confirm reprocessing as desired choice (WPI Report, para. 8.32). Tolstoy actually said reprocessing en-

tails waste disposal problems at a time when 'the disposal problem is most unlikely to have been solved'.

2. BNFL accepted reprocessing was part of EEC policy. Mr Justice Parker dismissed the implications by saying 'it is not for me to go into EEC policy' (Transcript, Day 64, p. 88).

3. Delays of two years in giving permission were accepted by BNFL and CEGB as not being greatly significant. The WI Report recommends a start on THORP 'without delay'.

4. Mr Justice Parker did not demonstrate that reprocessing would be justified by electricity demand once latter is disassociated from electricity use for heating.

(Source: Windscale Appeal, press release, April 1978).

New Ecologist

1. Mr Justice Parker has pre-empted outcome of fast reactor hearings despite claim to contrary. Unclear what the recovered Pu is for if not for fast reactors.

2. Mr Justice Parker took BNFL's word on corrosion of stainless steel clad elements, but no research had been done on the subject and no doubts about storage had been expressed prior to the Inquiry.

3. Vitrification is surrounded by doubts – internal heat can fracture the glass and sea-bed disposal has unknown consequences.

4. Periods necessary for isolation of products such as Caesium 134 exceed forecastable time-horizons.

5. WI Report unduly dismissive of evidence that linear-dose hypothesis wrong in respect of low-level radiation releases.

6. ICRP maximum permissible radiation levels have been determined by laxer criteria over the years.

7. BNFL's past record on aqueous discharges and releases of C134 and C137 give no cause

for faith in promises about future discharges.

8. Mr Justice Parker exaggerates value of reprocessed fuel for thermal plant.

9. All other reprocessing plants elsewhere have closed down or have escalated in cost by orders of magnitude.

(From *The Ecologist*, 'Reprocessing the Truth', May 1978.)

Lawyers' Ecology Group

1. Report failed even to summarise some objectors' arguments. Some opinions were expressed without explanation, thus misconstruing the Inspector's information role.

(Source: Letter by LEG to Peter Shore, 19 May 1978.)

Members of Parliament (House of Commons)

Leo Abse

1. WI Report's findings contrary to Carter NPT initiative.

2. Need to wait for findings of INFCEP.

Robin Cook

1. Witness Taylor misrepresented as supporting BNFL's capacity to meet any standard. What was said was that BNFL's expertise was such that they should design to a higher standard than they agreed to.

2. Witness Taylor misquoted on relationship between economic growth and nuclear power, Taylor not being an economist, a fact which Mr Justice Parker noted.

3. Mr Justice Parker quotes Ford/MITRE report on hazards of conventional fuels, but not the chapter on reprocessing which it states is disadvantageous and unnecessary.

4. The routine radiation discharge standards which are dramatically different for proposed US and German plant go unmentioned in WI Report.

5. Mr Justice Parker's reference to the desirability of spiking is unrelated to fact that foreign contracts contain no allowance for it.

6. Even if spiking is accepted, it implies knowledge of foreign plant specifications and requires, for Japan, construction of an LWR here. It is also costly.

7. NRC of USA and letter to Foreign Secretary (December 1977) by Joseph Nye states that spiking will *not* prevent a 'threshold' country from obtaining Pu.

8. Mr Justice Parker's interpretation of NPT is inconsistent with existence of Nuclear Supplies Group, of which the UK is a member, and which aims to restrict 'the transfer of sensitive nuclear technology'.

9. Mr Justice Parker was in possession of Nye's letter to Foreign Secretary which explicitly opposes further reprocessing plant. The Foreign Office's reply to Nye states that the letter was brought to the notice of the Inspector.

David Penhaligon

1. Wait for INFCEP.
2. WI Report inconsistent with US policy.
3. Vitrification unproven.
4. The security aspects of THORP are too clouded in secrecy.

Frank Hooley

1. Nye's letter (see Cook's objections) states that BNFL may not count on permission to transfer fuel for foreign contracts. It also explicitly refers to 'incorrect' conclusions in Mr Justice Parker's assessment that US will consent to retransfers after reprocessing.
2. Wait for INFCEP.

Appendix 3
International Linkages in Nuclear Fuel Services

The owners of THORP are British Nuclear Fuels Limited. Chapter 6 showed that THORP will be designed to reprocess spent oxide fuel from UK advanced gas-cooled reactors and oxide fuel from overseas reactors. In 1978 BNFL had contracts with the following countries for reprocessing oxide fuels: Canada, Germany, Italy, Japan, Netherlands, Spain, Sweden and Switzerland. The possibility of further contracts was being discussed with a number of the European countries with nuclear power programmes.

It is helpful to see BNFL in its international context. Apart from contracts with other countries for reprocessing fuels, BNFL owns part shares in seven other companies and is the 100 per cent owner of BNFL Enrichment Limited. Its part ownership brings it into contact with numerous European companies. The precise linkages in 1978 are shown in Figure A3.1.

Some additional information on URENCO and COGEMA is helpful since both organisations are involved in international dealings which have attracted some attention.

URENCO/CENTEC

According to BNFL's Seventh Annual Report (1977/78) URENCO has existing orders worth about £1000m. This new business would therefore require further extensions to the enrichment plant capacity at Capenhurst and Almelo (Netherlands) during the period up to the mid-1980s. The French are enlarging their uranium enrichment capacity also. Their Eurodif plant will probably be in operation by 1979, in which case URENCO and Eurodif together will constitute one third of the Western enrichment capability in 1985. Belgium, Spain, Italy and Iran are participating in the Eurodif development but the French have the majority interest. Conflict occurred between the members of URENCO over the contract to supply Brazil with enriched uranium. The Dutch anti-nuclear movement concentrated on lobbying their government to prevent the export of enriched uranium to Brazil, which is a non-signatory of the Non-Proliferation Treaty. A BNFL spokesman was reported in the *Guardian* on 26 June 1978 as saying that there are many safeguards against the misuse of civil reactor grade fuel and that enriched uranium would be supplied to Brazil under strict IAEA safeguards as well as those of Euratom.

West Germany wants enriched uranium sent to Brazil to fuel the eight reactors they are supplying along with their own enrichment plant and perhaps a reprocessing facility.

Matters were complicated in March 1978 by a fire which broke out at the Brazilian Westinghouse reactor, which was nearing completion. Engineers working on the site refused to sign a document saying that the plant was safe. This document was to have been presented to West Germany to stem fears in the URENCO consortium about supplying Brazil with enriched uranium.

Even before the news of the fire URENCO had decided not to sell enriched uranium to Brazil. In fact URENCO is unwilling to sell enriched uranium to any country which is not prepared to accept full safeguard measures.

In May 1978, the Dutch were still trying to reach a compromise with Britain and West Germany.

The Germans have decided to build their own enrichment plant (Urenco Deutschland) which would be within the Urenco Consortium.

COGEMA

COGEMA runs the French reprocessing plants in present use and is responsible for those either under construction now or planned for the future. France had a proposed contract with Pakistan to provide them with reprocessing facilities. This agreement was made in 1976. It was reported in the magazine *Science* in January 1977 that the French had renounced future exports of reprocessing technology in a formal statement, but in March 1978 the *Guardian* stated that there was no decision as yet on the export of reprocessing facilities to Pakistan. Finally it was reported in *The Times* on 24 August 1978 that the French had decided not to sell nuclear reprocessing plant to Pakistan under the terms of the 1976 agreement. Some leeway has been retained, however, by saying that they awaited further modifications made to the agreement.

Meanwhile COGEMA has contracts with a number of countries to reprocess more than 6000 tons of their spent fuel. Of this, 1705 tons is coming from West German reactors from 1980 onwards. Some of the rest will be from Japan and Sweden. Further contracts were reported as under discussion with Belgium, Austria, Switzerland, the Netherlands and Finland.

Notes on Figure A3.1

UKAEA	UK Atomic Energy Authority. Research and development on all aspects of nuclear power in support of the UK nuclear industry and under contract to overseas customers.
GEC	General Electric Company. Manufacturer of a wide range of equipment from powers generators to electronics.
British Nuclear Associates	A holding group representing various private commercial organisations in the nuclear industry, namely:
Taylor Woodrow	Building and civil engineering group. Designed cylindrical type pressure vessels for AGRs.
Head Wrightson	Design and contracting. Supplies and erects boiler shield walls, reactor core, guide tubes and other components for AGR power stations.
Sir Robert McAlpine	Building and civil engineers.
Strachan & Henshaw	Designs and manufactures nuclear reactor refuelling machinery including that for handling of nuclear fuel
Clark Chapman/ John Thomson	Heavy engineering, steam generators and pressure vessels. Designs and manufacturers mechanical handling plants.
Combustibili Nucleari	Manufactures and supplies fuel and fuel components for nuclear power stations. Located in Italy.
CENTEC	Co-ordinates the development in the laboratories in the three countries (Netherlands, Britain and West Germany) of the tripartite àgreement of plant for the enrichment of uranium by the centrifuge process.
URENCO LIMITED	Markets uranium enrichment services for the partners in the Tripartite Agreement. Located at Marlow, Bucks.
	Production is at present undertaken by the partnerships Urenco UK (at Capenhurst, Cheshire) and Urenco Nederland (at Almelo in the Netherlands). In each of these partnerships the local partner has a controlling share and the other two partners, minority shares.

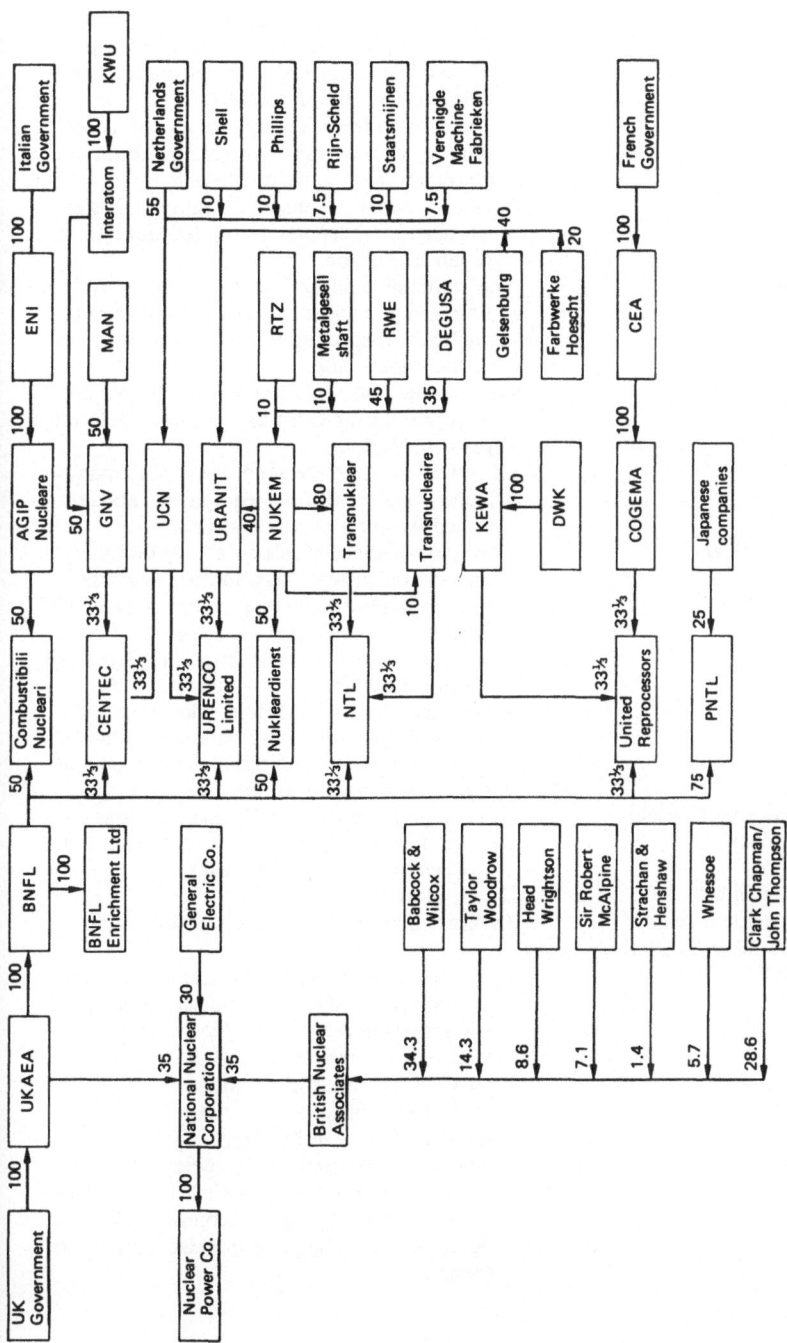

FIG. A3.1 **BNFL links with nuclear industry (numbers represent percentage shareholding)**

NTL	Nuclear Transport Limited, located at Risley, Cheshire. Provides services for the transport of irradiated nuclear fuel elements from oxide-fuelled power reactors.
Nukleardienst	Set up to provide nuclear fuel services for AGR-type reactors which might be built in Germany. Not active at present.
PNTL	Pacific Nuclear Transport Limited, Risley, Cheshire. Transports irradiated fuel from Japan to Europe.
Agip Nucleare	The national nuclear fuel company of Italy. Engaged in nuclear fuel cycle activities ranging from prospecting and refining of uranium to fuel fabrication.
GNV	Gessellschaft für Nukleare Verfahrenstechnik. Responsible for German national R&D on gas centrifuges.
UCN	Ultra-Centrifuge Nederland. Formed to manufacture gas centrifuge equipment and enriched uranium plant construction.
Transnuklear	Transport of nuclear fuels, located in Germany.
Transnucleaire	Transport of nuclear fuels, located in France.
COGEMA	A French state-owned company involved in all aspects of the nuclear fuel cycle.
Nukem	Nuclear Chemie und Mettalurgi. Nuclear fuel manufacture and nuclear fuel cycle services.
MAN	Maschinenfabrik Augsburg–Nürnberg. Research development and component manufacture for power and research reactors and high speed gas centrifuges.
Farbwerke Hoescht	A leading German chemical manufacturer.
Rijnscheld	Reactor Pressure vessels; heat exchangers instrumentation; transport casks.
Staatsmijnen	Dutch State Mines.
Verenigde machine fabrieken/STORK	Turbines and other large machanical parts.
Metalgesellschaft	A metal company selling bismuth, cadmium, lead and tellurium to the nuclear industry.
RWE	Rheinisch–Westphaelisches Elekrizitaetswerk. Germany's largest electricity utility.
Degussa	Control rods and special thermocouples for nuclear reactors as well as industrial furnaces for nuclear fuel technology.
DWK	Deutsch Gessellschaft für Wiederaufarbettung von Kernbrennstoffen. Aim is to form technical, economic and legal preconditions for the operation of a large reprocessing plant in Germany.

Appendix 4
Attitudes of other Countries towards Reprocessing

(a) The European Economic Community
The EEC is keen to promote reprocessing as a joint European venture (EEC *Bulletin*, August, 1977). Ideally the Commission would like to keep reprocessing facilities to a minimum but would wish to see each plant large enough to accommodate the reprocessing needs of member countries on economic terms. It is the opinion of the Commission that the institution of regional reprocessing centres would also reduce costs, safety hazards and threats of theft or sabotage. In conjunction with their approval of reprocessing the EEC propose to introduce a twelve-year plan (1978–90) of joint action to investigate all the problems caused by radioactive waste. This will include measures to facilitate the introduction of an EEC network of storage sites (EEC *Bulletin*, August 1977).

(b) Reprocessing: the relationship between the USA and Europe
The nuclear fuel cycle policy of the USA is likely to have some influence on European attitudes towards reprocessing. Fundamentally, President Carter considered that this stage of the fuel cycle implies a growth in trade of plutonium throughout the world. Traffic in plutonium-producing technology would contribute to the acquisition of atomic bomb material by non-nuclear weapon states.[1] In April 1977 President Carter said:

> The US is deeply concerned about the consequences for all nations of a further spread of nuclear weapons or explosive capabilities. We believe that these risks would be vastly increased by the further spread of sensitive technologies which entail direct access to plutonium, highly enriched uranium or other weapons-useable material. (Statement by the President on nuclear power, 7 April 1977)

This idea bears a resemblance to that considered by the EEC. The 1970

279

Non-Proliferation Treaty was aimed at promoting international co-operation in fuel cycle services by providing a mutual undertaking between non-nuclear and nuclear-weapon states. In return for assistance given by nuclear-weapon states in the development of nuclear power for peaceful purposes, non-nuclear-weapon states commit themselves not to procure atomic weapons. The Americans want the European nuclear-power states to support this objective. Currently the Carter administration has attempted to pressurise European countries into ceasing trade in what the USA considers to be potentially dangerous nuclear fuel cycle facilities. France and West Germany have been forced to reconsider selling reprocessing plant to Pakistan and Brazil respectively. In Britain BNFL have signed a contract in 1978 with Japanese companies to reprocess 1600 tons of spent fuel but the Japanese have yet to obtain American permission to carry out the terms of the contract. Japan is affected by the USA's moritorium on reprocessing, which not only impedes plant development in America but also prevents other countries from exporting spent fuel for processing which had its origins in enriched uranium supplied by the USA. Consequently Japan must apply to the USA for permission to export spent fuel to Britain and France for reprocessing. At present America judges such petitions on a case-by-case basis. The Japanese have four years in which to persuade the USA to give this permission, some consignments having already been permitted.

In the meantime the Americans have managed to convince the nations of Europe that they should take part in a two-year International Nuclear Fuel Cycle Evaluation Programme (INFCE). Its terms of reference include the investigation of alternative (more proliferation-proof) fuel cycles and measures to facilitate the development of international storage sites. The INFCE also harks back to the theme that countries with a non-proliferation stance should be assured of access to nuclear fuel supplies.

Although in the USA reprocessing has come to a halt and the fast reactor programme is actively discouraged the development of uranium enrichment plant is to go on apace. The USA hopes to fulfil the world's need for enriched uranium so that other nations will not consider the acquisition of reprocessing facilities a desirable proposition. Meanwhile in Europe by the mid-1980s URENCO and Eurodif[2] will be providing enough enriched uranium to satisfy 80 per cent of non-American requirements (*Financial Times* European Energy Report No. 7, 1978). This should result in a significant decline of American influence over European nuclear policy.

(c) France

The French possess one of the fastest expanding nuclear power programmes in Europe, the impetus to which was given by the massive increase in oil prices in 1974. At present the target for nuclear power is to provide France with 70 per cent of its electricity and 25 per cent of its energy by 1985 (Schwarz, 1978).

Through COGEMA the French are members of United Reprocessors (see Appendix 3). Their uranium oxide reprocessing facility is sited at a La Hague. Contracts have been entered into with Japan to reprocess 1600 tons of spent fuel, Sweden (620 tons) and recently with West Germany (1705 tons). This totals 3925 tons of foreign irradiated fuel requiring reprocessing, exclusive of an increasing domestic demand. The 800 t/y UP$_2$ (see Table A4.1) facility will be taken up with domestic reprocessing needs by 1985–6. Hence a new reprocessing plant UP$_3$ on the same site is to be built and is planned to come into operation in 1985. Initially this new plant will have a capacity of 800 t/y but this will rise to 1600 t/y, thereby allowing France to increase the amount of foreign spent fuel she is able to reprocess. At present negotiations are in progress with Belgium, Austria, Switzerland, the Netherlands and Finland (*Times*, 14 April 1978).

The French are involved in the mining of uranium in Niger: enrichment through the Eurodif plant, the manufacture of Nuclear Power stations for domestic needs and the export market, reprocessing and waste disposal. They are now embarking on a fast reactor programme at Creys Malville and will be co-operating with West Germany on the development of this technology (Dewhirst, 1977).

(d) West Germany

As in France, when oil prices increased in 1974 West Germany drew up an ambitious plan for the enlargement of its nuclear power programme. This source of power was to supply them with 45 per cent of their energy requirements by the mid-1980s. The plan has since been revised and instead of 45,000–50,000 MW of installed nuclear capacity being available by 1985 the figure is now estimated to the nearer 25,000 MW. In conjunction with this deceleration recent Court decisions have forbidden further building of nuclear plant until safe methods for the final disposal of radioactive wastes have been found. Currently West Germany possesses thirteen power stations which have an overall capacity of 6400 MW,[3] producing about 11 per cent of their total electricity output (Buschschluter, 1978). As yet West Germany only has a small uranium oxide reprocessing plant of 40 t/y sited at Karlsruhe but a 1500 t/y facility was expected to be in operation by the late 1980s. This facility would probably be constructed by KEWA (see Figure A3.1), the West German partner in United Reprocessors. At the time of writing, this plant has been postponed indefinitely.

(e) Canada

Unlike the USA, Great Britain, France and West Germany, Canada has never concluded that reprocessing is a necessary stage in the nuclear fuel cycle. The CANDU reactor, which utilises heavy water as the moderator and natural uranium as the fuel, was developed in Canada. It has low refuelling costs achieved by 'the use of natural uranium, reduction of restrictive handling requirements and standardised uncomplicated fuel design' (Boyd, 1974). This, coupled with the fact that Canada's reserves of uranium ore amount to almost 25 per cent of the world's current total, means that reprocessing is not an economic necessity at the moment. When Canada's indigenous uranium reserves begin to wane the CANDU reactor can be modified to operate on thorium fuel. Thorium is estimated to be over three times as abundant worldwide as uranium (Robertson, 1978). Simultaneous investigations are under way into methods of permanent disposal of unreprocessed irradiated fuel and economic analysis of fuel recycling. Nevertheless, at present 'the overall objective of the waste management program is to provide safe storage of spent fuel until society determines whether the large amount of latent energy it contains should be exploited through fuel recycling' (Robertson, 1978).

NOTES

1. Dickson, David (1978), 'U.S. Congressional Committee rejects Carter Compromise on Fast Reactor Programme', *Nature*, April 1978. In this article Mr Dickson states that President Carter, finding his anti-proliferation of nuclear weapons stand has had little effect upon European policy, has since altered the direction of his challenge. He now wishes to debate the size and shape of future energy demand.
2. The Eurodif plant went on stream in February 1978. The full capacity of the plant is expected to be 10.8 million Separative Work Units, which is equivalent to 2670 t of uranium enriched to 3.15 per cent.
3. On 25 November *The Times* stated that present German installed capacity was 7400 MW.
4. This table is a modification of Table 2: World Reprocessing Plants, in Franklin, N. L. *Irradiated Fuel Cycle* (BNFL, 1975).

REFERENCES

EEC (1977)   'The Community's Nuclear Strategy'
                            Bulletin of the European Community,
                            August 1977.
Hazelhurst, P. (1978)   'U.K. signs £500m nuclear deal with
                            Japanese but US approval is still needed',
                            Times, 25 May.
Financial Times (1978)   *European Energy Report*, Issue No. 7.
Schwarz, W. (1977)   'Doubts cast shadow over France's Nu-
                            clear Future' *Guardian*, 29 December.
The Times (1978)   'French Reprocessing', 14 April.
Dewhirst, P. (1977)   'France, W. Germany to co-operate on
                            Fast Breeder Nuclear Reactors' *World
                            Environment Report*, 15 August
Buschschluter, S. (1978)   'When the demos have to stop', *Guardian*,
                            24 April.
Boyd, F. C. (1974)   'Nuclear Power in Canada: a different
                            approach', *Energy Policy*, June.
Robertson, J. A. L.   'The CANDU reactor system: an Appro-
                            priate Technology?', *Science*, vol. 199.
BNFL Document 7   Remarks of the President of the United
presented to the   States of America on Nuclear Power
Windscale Inquiry.   Policy and Question and Answer Session
                            (Office of the White House Press
                            Secretary, 7 April 1977).

TABLE A4.1   Reprocessing capacity in France, Great Britain, USA and West Germany[4]

Country	Plant	Location	Owner	Capacity	Operational Date
France	UP1	Marcoule	COGEMA	900–1200 t/y (metal fuel)	1958
(Plant formally run by C.E.A. Inherited by COGEMA 1976)	UP2	La Hague	COGEMA	1000 t/y (metal fuel) 400 t/y – 800 t/y (UO₂)	1966 1976 UO₂ to replace metal fuel plant. Capacity rising to 800 t/y by 1980s.
	UP3	La Hague	COGEMA	800 t/y – 1600 t/y (UO₂)	1985
Great Britain		Windscale	B.N.F.L.	2000–2500 tU/y (inc. 400 t/y oxide) 1200 t/y (UO₂)	1964 Head-end treatment plant operational again in near future 1980s.
U.S.A.	NFS	West Valley, N.Y.	Getty Oil Skelly Oil	300 t/y (UO₂) 40 Kg/d highly enriched fuels	1966 Not operating. Owners feel plant has no economic future.
	MFRP	Morris, Ill.	General Electric	300 t/y (UO₂)	Not operating. Future of plant undecided.
	AGNS	Barnwell, S.C.	Allied Chemical Co. General Atomic Co.	1500 t/y (UO₂)	1976 Not operating. Future of plant undecided.
West Germany	WAK	Karlsruhe Gorleben	GWK KEWA	40 t/y (UO₂) 1500 t/y (UO₂)	1971 1980s (but deferred indefinitely).

Appendix 5
Existing Commissions

(a) THE ENERGY COMMISSION

Membership
Secretary of State for Energy (Chair)
Minister of State, Scottish Office

Representatives of Energy Industries

1. Mr D. R. Berridge	Chairman, South of Scotland Electricity Board
2. Sir D. Ezra	Chairman, National Coal Board
3. Sir John Hill	Chairman, Atomic Energy Authority
4. Lord Kearton	Chairman, British National Oil Corporation
5. Sir D. Rooke	Chairman, British Gas Corporation
6. Sir F. L. Tombs	Chairman, Electricity Council

Representatives of Trades Union Council Fuel and Power Industries Committee

7. Mr F. A. Baker	National Industries Officer of the National Union of General and Municipal Workers
8. Mr R. Birch	Member of the Executive Council of the Allied Union of Engineering Workers
9. Mr F. Chapple	General Secretary, Electrical Electronic Telecommunications and Plumbing Union
10. Mr F. A. Drain	General Secretary, National and Local Government Officers' Association
11. Mr J. Gormley	General Secretary, National Union of Mineworkers
12. Mr D. Lea	Secretary, TUC Fuel and Power Industries Committee
13. Mr C. Unwin	Deputy Secretary, Transport and General Workers' Union

Non-producer Members

14. Mr M. C. J. Barnes	Chairman, Electricity Consumers' Council
15. Mr R. T. Carlisle	Managing Director, Babcock & Wilcox Ltd
16. Lord B. Flowers	Rector of Imperial College, London
17. Professor Sir W. Hawthorne	Master of Churchill College, Cambridge
18. Mr R. Lawrence	Vice-Chairman, British Railways Board
19. Baroness McLeod of Borve	National Gas Consumers Council
20. Mr E. C. Sayers	Chairman, CBI Energy Policy Committee; Chairman, Duport Ltd

(b) THE COMMISSION ON ENERGY AND THE ENVIRONMENT

Membership

1. Chairman: Lord B. Flowers	Rector of Imperial College, London
2. Professor T. J. Chandler	Master of Birbeck College, London
3. Sir A. H. Chilver	Vice-Chancellor, Cranfield Institute of Technology
4. Dr J. Collingwood	Director of Unilever, Fellow of University College, London
5. Mr A. Derbyshire	Partner, Robert Matthew Johnson-Marshall; Part-time Member, CEGB
6. Professor Sir R. Doll	Regius Professor of Medicine, University of Oxford
7. Professor Sir W. Hawthorne	Master, Churchill College, Cambridge
8. Professor F. Holliday	Professor of Zoology, Aberdeen University
9. Dr A. Pearce	Chairman, Esso (UK)
10. Mr M. Posner	Fellow, Pembroke College, Cambridge; Chairman, SSRC
11. Sir F. Tombs	Chairman, Electricity Council
12. Mrs V. Milligan	Vice-Chairman, East Wales Institution of Electrical Engineers
13. Mrs N. McIntosh	Gas Consumer Council

14. Mr R. Bottini	Ex-General Secretary, NUAW; Member, General Council, TUC
15. Mr G. McGuire	Vice-Chairman, Ramblers Association; Member, Countryside Commission
16. Mr D. E. T. Williams	Reader in Public Law, University of Cambridge

(c) THE ROYAL COMMISSION ON ENVIRONMENTAL POLLUTION

The Royal Commission on Environmental Pollution was originally appointed in February 1970. Its terms of reference are 'to advise on matters, both national and international, concerning the pollution of the environment; on the adequacy of research in this field, and the future possibilities of danger to the environment'.

1. Professor Sir Hans L. Kornberg	Sir William Dunn, Professor of Biochemistry, University of Cambridge, and Fellow of Christ's College
2. The Marchioness of Anglesey	Chairman of the Welsh Arts Council and Member of the Arts Council of Great Britain; Deputy Chairman of the Prince of Wales' Committee and member of the IBA.
3. Sir A. H. Chilver	Vice-Chancellor, Cranfield Institute of Technology, and Chairman of the Computer Board for Universities and Research Councils
4. Dr J. G. Collingwood	Member of the Food Standards Committee and of the Council of the University of Aston; Fellow of University College London
5. Professor Sir R. Doll	Regius Professor of Medicine, Oxford University
6. Mr R. A. Grantham	General Secretary of the Association of Professional, Executive, Clerical and Computer Staff
7. Professor P. D. Henderson	Professor of Political Economy, University College London
8. Professor P. J. Lindop	Professor of Radiation Biology, University of London; Head of Depart-

	ment of Radiobiology at the Medical College of St Bartholomew's Hospital
9. Mr J. R. Maddox	Director of the Nuffield Foundation; writer and broadcaster
10. Professor J. M. Mitchison	Professor of Zoology, University of Edinburgh
11. Professor R. E. Nicoll	Professor of Urban and Regional Planning, Strathclyde University; Director of Glasgow Chamber of Commerce and a member of the Scottish Council (Development and Industry)
12. Professor T. R. E. Southwood	Professor of Zoology and Applied Entomology, University of London; Chairman of the Division of Life Sciences; Head of Department of Zoology and Applied Entomology and Director of Field Station, Imperial College
13. Mr R. E. Thornton	Farmer; member of the Thames Water Authority, Chairman of the Regional Land Drainage Committee, member of the Ministry of Agriculture Regional Board and an official of the Country Landowners Association
14. Sir R. Verney	Chairman of the Forestry Commission's Committee for England; Chairman of the Government Advisory Committee on Aggregates
15. The Baroness White	Chairman of the Land Authority for Wales; member of the British Waterways Board; Chairman of the Advisory Committee on Oil Pollution of the Sea; member of the Waste Management Advisory Council
16. Mr D. G. T. Williams	Reader in Public Law, University of Cambridge; Fellow of Emmanuel College; member of the Council on Tribunals and of the Clean Air Council

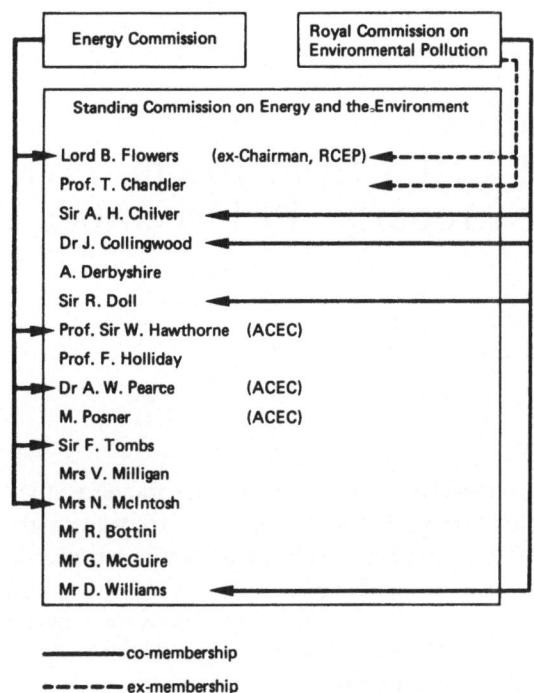

FIG A5.1 **Linkages between EC, RCEP and CEE (August 1978)**

Appendix 6
Text of Letter from Joseph Nye to Patrick Moberly, 19 December 1977

Patrick Moberly, Esquire
Foreign and Commonwealth Office
London

Dear Patrick:

Our attention has been called to a misunderstanding of U.S. policy in the Windscale hearings. Let me try to set the record straight. Our view that it is not wise to build more solvent extraction reprocessing plants at this time has not changed since we discussed the question during consultations last April. The testimony draws an unwarranted conclusion from accommodations that we have made to avoid disruption of existing nuclear power programs.

On the broad question of whether to build a new 1200 MT per year reprocessing facility at Windscale, we of course recognize that this is a matter for your own Government to decide. As you know, President Carter has decided to defer indefinitely commercial reprocessing in the United States. This decision was based not only on the paramount need to seek ways of reducing the associated proliferation risks, but also on the conclusion that reprocessing was not urgently needed in the near term and was not essential to – and could in fact exacerbate – the problems of waste disposal.

While it has been argued that there are potential non-proliferation advantages of reprocessing services being supplied by a few supplier states, this solution entails three major difficulties: (1) If such services involve a requirement to return the separated plutonium directly to the customers, the proliferation gain is illusory. The plutonium will be moving in international commerce, and the customer will be getting it without going to the trouble, delay and expense of developing its own reprocessing capability. (2) Denial of the plutonium to the customer, on the other hand, may make the customer unwilling to enter the deal, add to the complaints that industrialized states are trying to deny to developing countries fuels and technology, and in the case of nuclear-weapons states, that they are now trying to extend the discriminatory features of the NPT to the civil nuclear field, and thus increase the likelihood that the customer will build its own reprocessing facility. (3) The 'compromise' of returning the plutonium only in the form of fabricated fuel elements seems unlikely to be acceptable to other industrialized countries, which have their own fabrication capability designed to meet their nuclear reactor fuel needs. And for threshold countries, it does not present much of an obstacle to obtaining the plutonium.

Accordingly, the United States is not prepared at this time to encourage weapons states to decide in favor of proceeding with new reprocessing plants. Thus, while we will continue to consider MB-10 requests on a case-by-case basis, and grant them in cases where there is a demonstrated need, such as inadequate spent fuel storage capacity, we cannot give any assurance that BNFL may count on MB-10s as a matter of course for feed for a new plant or in support of long-term reprocessing commitments that it may enter into.

Finally, as we have noted in each of the MB-10 requests involving transfers for reprocessing that we have granted in the past two years, the granting of such requests would be subject to our right of consent over the disposition of the nuclear materials resulting from such reprocessing. The enclosed excerpt from the Windscale hearings incorrectly concludes that we are prepared to consent now to retransfers of such products after reprocessing which will not take place for at least 13 years. The enclosed Department of Energy letter from which this conclusion was drawn stated that 'in accordance with applicable agreements for cooperation, such transfers would, at that time, have to be approved by the Government of the United States.'

I hope that this clarification of our policy is helpful. I am sure that you agree that we should seek to avoid any future misunderstandings that might arise from a lack of clarity now.

Sincerely,

Joseph S. Nye
Deputy to the Under Secretary

Author Index

Subject Index